DEATH IN A SMALL PACKAGE

JOHNS HOPKINS BIOGRAPHIES OF DISEASE
Charles E. Rosenberg, Series Editor

DEATH IN
A SMALL
PACKAGE

❋ ❋ ❋

A Short History of Anthrax

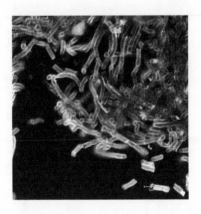

Susan D. Jones

THE JOHNS HOPKINS UNIVERSITY PRESS
Baltimore

The Johns Hopkins University Press
2715 North Charles Street
Baltimore, Maryland 21218-4363
www.press.jhu.edu

Library of Congress Cataloging-in-Publication Data

Jones, Susan D., 1964–
Death in a small package : a short history of
anthrax / Susan D. Jones.
p. ; cm. — (Johns Hopkins biographies of disease)
Includes bibliographical references and index.
ISBN-13: 978-0-8018-9696-5 (hardcover : alk. paper)
ISBN-10: 0-8018-9696-7 (hardcover : alk. paper)
1. Anthrax—History. 2. Biological weapons—
History. I. Title. II. Series: Johns Hopkins
biographies of disease.
[DNLM: 1. Anthrax—history. 2. Biological warfare agents—
history. WC 305 J79d 2010]
QR201.A6J66 2010
616.9′56—dc22   2009052700

A catalog record for this book is available from the British Library.

*Special discounts are available for bulk purchases of this book. For more
information, please contact Special Sales at 410-516-6936 or
specialsales@press.jhu.edu.*

The Johns Hopkins University Press uses environmentally friendly
book materials, including recycled text paper that is composed of
at least 30 percent post-consumer waste, whenever possible. All of
our book papers are acid-free, and our jackets and covers are
printed on paper with recycled content.

For Phil,
who wrote many books
while I finished this one

But now we come to another kind of war, war waged not against germs but with germs against men, animals, and plants—BW [biological weapons] . . . What distinguishes a potential BW agent from just any germ? . . . Infectivity; casualty effectiveness; availability; resistance; means of transmission; specific immunization; therapy; detection; and retroactivity.

Theodor Rosebury, *Peace or Pestilence: Biological Warfare and How to Avoid It,* 1949

CONTENTS

Disease is a fundamental aspect of the human condition. Ancient bones tell us that pathological processes are older than humankind's written records, and sickness and death still confound our generation's technological pride. We have not banished pain, disability, or the fear of death, even if we die on average at older ages, of chronic and not acute ills, in hospital or hospice beds and not in our own homes. Disease is something men and women feel. It is experienced in our bodies—but also in our minds and emotions. Disease demands explanation; we think about it, and we think with it. Why have I become ill? And why now? How is my body different in sickness from its quiet and unobtrusive functioning in health? Why in times of epidemic has a whole community been scourged?

Answers to such timeless questions necessarily mirror and incorporate time- and place-specific ideas, social assumptions, and technological options. In this sense, disease has always been a social and linguistic, a cultural as well as biological, entity. In the Hippocratic era, physicians—and we have always had them with us—were limited to the evidence of their senses in diagnosing a fever, an abnormal discharge, or seizures. Classical notions of the somatic basis for such felt and visible symptoms necessarily reflected and incorporated contemporary philosophical and physiological notions, a speculative world of disordered humors, "breath," and pathogenic local environments. Today we can call for understanding on a variety of scientific insights and an armory of diagnostic and therapeutic practices—tools that allow us to diagnose ailments unfelt by patients and imperceptible to the doctor's senses. In the past century, disease has become increasingly a bureaucratic phenomenon as well—as sickness has been defined and in that sense constituted by formal disease classifications, treatment protocols, and laboratory thresholds.

Sickness is also linked to climatic and geographic factors. How and where we live and how we distribute our resources all contribute to the incidence of disease. For example, ailments such as typhus fever, plague, malaria, dengue, and yellow fever reflect specific environments that we have shared with our insect contemporaries. But humankind's physical circumstances are determined in part by culture, and especially by agricultural practice in the millennia before the growth of cities and industry. The history of anthrax, for example, as readers of this book will see, reflects shifting relationships among animals, soils, and human behaviors. Environment, demography, ideas, and applied medical knowledge all interact to create particular distributions of disease at particular moments in time. The contemporary ecology of sickness in the developed world is marked, for example, by the dominance of chronic and degenerative illness—ailments of the cardiovascular system, of the kidneys, and cancer. But this has not always been so.

Disease is thus historically as well as ecologically and biologically specific. Or perhaps I should say that every disease has a unique past. Once discerned and named, every disease claims its own history. At one level, biology creates that idiosyncratic identity. Symptoms and epidemiology as well as generation-specific cultural values and scientific understanding shape responses to illness. Some writers may have romanticized tuberculosis—think of Greta Garbo as Camille—but as the distinguished medical historian Owsei Temkin noted dryly, no one had ever thought to romanticize dysentery. Tuberculosis was pervasive in nineteenth-century Europe and North America and killed far more people than cholera did but never mobilized the same widespread and policy-shifting anxiety as cholera. Unlike tuberculosis, cholera killed quickly and dramatically and was never accepted as a condition of life in Europe and North America. Its episodic visits were anticipated with fear. Sporadic cases of influenza are normally invisible, indistinguishable among a variety of respiratory infections;

waves of epidemic flu are all too visible. Syphilis and other sexually transmitted diseases, to cite another example, have had a peculiar and morally inflected attitudinal history. Some diseases, such as small-pox or malaria, have a long history; others, like AIDS, a rather short one. Some have flourished under modern conditions; others seem to reflect the realities of an earlier and economically less developed world.

These arguments constitute the logic motivating and underlying the Johns Hopkins Biographies of Disease. *Biography* implies a coherent identity, a chronology and a narrative—a movement in and through time. Once inscribed by name in our collective understanding of medicine, each disease entity becomes a part of that collective understanding and thus inevitably shapes the way in which individual men and women think about their own felt symptoms and prospects for future health. Each historically visible entity—each disease—has a distinct history, even if that history is not always defined in terms familiar to twenty-first-century physicians. Dropsy and Bright's disease are no longer terms of everyday clinical practice, but they are integral to the history of chronic kidney disease. Nor do we speak of essential, continued, and remittent fevers as categories in our classifications of disease, even though they played an important role in the history of medical ideas.

In fact, the very term *medical ideas* seems a trifle constrained, anachronistic in a sense. A generation of historical research and re-thinking has questioned the rigidity of the unquestioned boundaries between the "medical" and everything else. On the one hand, we have seen the laboratory and basic sciences redefine our understandings of those biological processes underlying health, recovery, and death and thus the options available to physicians in their clinical encounters. We have also been made aware of how a variety of factors—income distribution, social policy, cultural practices, climate, and physical environment—can shape morbidity and mortality with or without inputs from changing therapeutics and public health practice. Medicine—when and how we get sick and what happens to us when we do—is in only one of its senses bounded by the ideas and practices of individuals who call themselves doctors.

The history of anthrax provides a dramatic, and timely, instance of the persistence, complexity, and tenacity of such realities. Today, most educated men and women who encounter the word *anthrax* think first of germ warfare, of spores capable of being inhaled and causing a quick and unpleasant death. To veterinarians and specialists in infectious disease, it is an animal disease that, like many others, has a persistent relationship to human beings. To anyone who has studied bacteriology or taken a course in the history of medicine, Robert Koch's work in elucidating the cause and transmission of this deadly ailment constitutes a landmark in the history of medicine. In a sense, the modern history of anthrax begins in the 1870s with those discoveries and the domestication of the anthrax bacillus in the laboratory's flasks and test tubes.

But the disease has had a much longer history. Susan Jones chronicles the observations linking a frightening acute sickness in domestic animals with specific soils and sporadic but linked skin and systemic ills in man. Contact between the Old World and the Americas spread the disease, and the resilient spores of anthrax gained global purchase as world commerce grew in scope and speed. As hides and fleece from places as remote as Central Asia arrived in the mills and factories of nineteenth-century Yorkshire, workers died of an acute pneumonia-like ailment that came to be called "woolsorters' disease," inhalation anthrax before the term existed. In the past half-century, anthrax has found a new soil in which to thrive—the world of bioweapons.

There is a moral here. Nature and culture are hard to disentangle. The anthrax organism is very much a part of nature, implacable and with its own biological character (altered now by two generations of zealous biowarriors in various laboratories). Yet its relationship to human beings is shaped and reshaped not only by its own genetics, but also by the actions of men and women who have, over the millennia, domesticated animals, shaped global commerce, and found new ways to contemplate war. For historians, the story of anthrax is peculiarly illuminating. Is it a part of economic or environmental history? Of political or medical history? Of historical demography? The moral is unavoidable. To trace the path of anthrax over time is to undermine the fragmenting boundaries that have shaped the writing of academic history in the past century.

Charles E. Rosenberg

I first encountered anthrax in the misty early morning of a fall day in the late 1970s, in rural Illinois. I was interested in a career in veterinary medicine, so I was shadowing the local vet, who was about to retire. Sitting in his dusty pickup truck among ropes, papers, and boxes of empty vaccine vials, I listened carefully as my mentor drove us around the county and pointed to the locations of old burial sites for cattle who had died of anthrax. "Don't ever allow digging there," he warned. "That'll expose the spores and you'll get another outbreak." The spores, I knew, contained the microorganisms that caused anthrax, a species of bacteria named *Bacillus anthracis*. I expressed surprise that the bacteria could lie dormant for so long and then still cause the disease, but he assured me that they could. The bacteria would kill cattle suddenly, and they could infect people, too; he knew a veterinarian who had almost died of anthrax. "Be careful and report any suspicious deaths to the state," he added. Although rare, outbreaks of anthrax were potentially dangerous unless handled correctly by the veterinarian.

Thirty years later, those Illinois burial sites have remained undisturbed, and no anthrax outbreaks have occurred there. Instead, we in the United States now think of anthrax primarily as a biological weapon, especially since the widely publicized events of 2001, when someone sent a series of letters containing *Bacillus anthracis* spores through the U.S. mail system. Tragically, five people who would not otherwise have had any contact with *B. anthracis* died of anthrax. Tens of thousands more in the eastern United States feared opening their mail, and the federal postal system and the U.S. Congress were paralyzed while investigators searched for more letters and workers decontaminated office buildings on Capitol Hill. With my background in veterinary medicine, I was struck by how different *this* anthrax was from the agricultural anthrax I remembered from that morning in the veterinarian's pick-up truck. The 2001 letters made the international news, which certainly would not have happened with an outbreak in cattle on one of those Illinois farms. Five people died in the 2001 letter attacks, but almost as many Americans had died of anthrax, unnoticed in the popular press, in the preceding years because of their proximity to livestock or their interest in playing imported drums (the leather stretched over the drumhead could be contaminated with *B. anthracis* spores).

We Americans have the luxury of thinking about anthrax primarily as a biological weapon precisely because our well-vaccinated livestock population (and people who might be exposed because of their occupations) rarely contracts the disease naturally. Parts of the rest of the world are not so fortunate. Anthrax has remained a naturally occurring disease in populations of cattle, goats, sheep, and other animals central to agriculture. Animals and people in Western African nations suffer a high incidence of anthrax, especially in areas of civil unrest and little veterinary or human preventive care. It is common in El Salvador (in horses, especially), Guatemala, Peru, and parts of Mexico, Honduras, Nicaragua, and Costa Rica. Around one hundred people die of anthrax each year in the central Asian nations of Azerbaijan, Kazakhstan, and Tajikistan, and the Russian name for anthrax (Siberian ulcer) continues to be justified by frequent outbreaks there. Livestock and people in Mongolia, parts of western China, and parts of India live with regular outbreaks of anthrax, often after desperately poor people have eaten meat from sick animals. The situation in some places is so dire that the human population is used as "sentinels," or indicators of how well ensconced the disease is in that local environment and livestock population.* Agricultural anthrax—"naturally" oc-

---

* My thanks to Martin Hugh-Jones. See Hugh-Jones, "1996–97 Global Anthrax Report," presented at the Third Annual Conference on Anthrax, September 7–11, 1998, Plymouth, England. Available online at www.vetmed.lsu.edu/whocc/globrept .htm (accessed July 27, 2009).

curring anthrax—is hardly a historical phenomenon in these areas, although it is not as severe a public health concern as malaria, tuberculosis, and HIV/AIDS, all of which infect more people each year. Nonetheless, anthrax is particularly deadly and particularly difficult to control in agricultural areas.

For me, the different ways in which we classify anthrax outbreaks are windows into the constantly changing ecological and cultural relationships between animals, the environment, and human society and culture. Because its causative bacillus lives in the soil, anthrax is a disease associated with certain soils in certain locations. But as animal populations have migrated or been relocated by human activities, the spores of *B. anthracis* have followed to new locations. The result of this "spore diaspora" has been the appearance of the disease in new forms, such as in factory workers who process the wool of infected animals. The biological characteristics of the bacillus have also appealed to people looking for microorganisms to make into weapons. Few organisms kill so efficiently, and even fewer present the stability in the environment that would enable humans to turn this killing capacity into a weapon, a desire stimulated by the arms races and global wars of the twentieth century.

Human cultural beliefs about anthrax have shaped all of these events. Over time, people have explained outbreaks of anthrax in very different ways that reflected the concerns of their time. To some, anthrax has been an animal disease caused by fate or the displeasure of the gods. To others, the appearance of an anthrax outbreak meant that human malice was at work (as in the case of using *B. anthracis* as a biological weapon). How we understand anthrax depends on the cumulative knowledge of generations of people who have raised and worked with livestock as well as the work of physicians, natural historians, and scientists.

Historians have pointed out, rightly, that the general fear of anthrax seems out of proportion to the small numbers of human and animal cases reported around the world each year. However, this fear has come about through humankind's historical experience with anthrax, much of which is described in this book. In large part, we fear anthrax because it can kill its victims suddenly and brutally. From the human point of view, anthrax is a deadly, even sadistic, disease. From anthrax's point of view (if it had one), killing the animal it infects is an absolute necessity to continue its life cycle. Unlike many pathogens, *Bacillus anthracis* cannot be directly transmitted between animals and must be acquired anew by each host from the environment. *Bacillus anthracis* develops and reproduces within the bloodstream of its host's body (not in the gut, from where it could easily escape to infect others). It must kill its host so that it may exit during after-death bleeding, move back into the environment, and infect a fresh host. Thus it has evolved the ability to make not just one but two lethal toxins in the bodies of its victims—all the better to ensure that the victim dies. The resulting manifestations of anthrax are severe, often lethal, and terrifying.

While I had long been collecting historical information and thinking about anthrax, the events of 2001 put the various identities of the disease in sharp relief—and pointed to larger questions about the human capacity to alter and control elements of the natural world in which we are embedded. Over the past two centuries, human activities have certainly altered outbreaks of anthrax, but we have not succeeded in controlling it. Human actions have, ironically, helped to make anthrax what it is: an entity that arises not only from nature but also from human intention, something that we can use to kill or injure each other. History shows us that we have the capacity to shape anthrax outbreaks, but only within the social and cultural frameworks of our time. Anthrax is probably not going away any time soon. We can hope that, in the twenty-first century, we will be able to explain anthrax outbreaks as the outcome of ecological interactions between people and their environment, and not automatically as a result of human malice.

DEATH IN A SMALL PACKAGE

Introduction

✻ ✻ ✻

Omnia mutantur, nos et mutamur in illis. (All things are changed, and we change with them.)

various, after Ovid

Life is an adventure in a world where nothing is static; . . . where man himself, like the sorcerer's apprentice, has set in motion forces that are potentially destructive and may someday escape his control.

René Dubos, *The Mirage of Health,* 1959

To preserve its own life cycle, anthrax must kill its human host. Yet humankind has incorporated it into our soils, laboratories, factories, communities, and lives. This paradox defines our relationship to anthrax and to the bacterium that we associate with the disease. It also drives the narrative in this book, which explores the historical development of our close relationship with anthrax.

Anthrax kills its victims dramatically and mysteriously. Once infected, an animal or human can feel fine one day—grazing on a pasture, say, or sitting with a slight headache through a meeting—and be dead the next. Depending on the part of the body initially infected, a human or animal anthrax victim has about a 20 to 95 percent chance of dying. While death may occur quickly, it is not pleasant. Malaise, headache, and fever usually progress to gastro-intestinal upset, a cough, then seizures and the signs of general organ failure. In many cases, boils arise on the skin, swelling tremendously and turning red, then ominously black in the center (thus the French name for the disease, *charbon*). The human victim is usually conscious, aware, and very anxious until soon before death; after death, blood oozes copiously from the bodily orifices. Upon autopsy, the blood inside the body looks dark, and organs such as the spleen are so damaged that they disintegrate, almost liquefied, in the examiner's hands (thus the German name, *Milzbrand*, or spleen fire). In the case of animals, the gory remains of anthrax's visitation must be disinfected or incinerated, deeply buried, and left undisturbed to prevent the disease's spread, with the location carefully remembered by subsequent generations.

I have begun this book in such a morbid fashion to illustrate an important historical point: the gruesome deaths attributed to anthrax throughout history have collectively given the disease a fearsome reputation. Indeed, the more people learned about anthrax, the more terrible it looked. The various forms of anthrax are now believed to be caused by specific strains of the bacterium *Bacillus anthracis*. *B. anthracis* microorganisms have an almost unique property among their peers: they manufacture two deadly toxins in the body of victims. One of these toxins is a neurotoxin, similar to that of tetanus; the other mimics the action of cholera toxin. Once the toxins have spread within a victim's organ systems, a case of anthrax usually ends in death.

Anthrax occurs mysteriously from time to time, and this is another factor in the disease's fearsome reputation. The "small package" in which Death arrives (the title of this book) refers to the hardy spores that the anthrax bacillus forms and that can survive for long periods outside a victim's body. The spores wait for the next victim to come along, making the disease seem to disappear for a while and then recur. For at least the past two thousand years, anthrax has existed in the pastoral regions of the world, remaining in the soil long after its initial victims moved on or died.

This microbial species has a ubiquitous environmental presence: most of the world's land mass contains anthrax spores of various strains (or genetic types). These days, the environmental, agricultural type of anthrax seldom makes headlines; rather, anthrax as a biological weapon dominates the news. Since 1940, anthrax has been weaponized in Japan, Britain, the United States, the Soviet Union, and many other places. It has been most famously deployed as a weapon of terror—causing relatively few deaths, but inciting fear and disrupting social processes—in Manchuria, Zimbabwe, the Soviet Union, and the United States. At the beginning of the twenty-first century, anthrax is thought of as a weapon of mass destruction, capable of killing humans and animals and destroying landscapes.

The major question that this book addresses is how a ubiquitous agricultural disease became a biological weapon. As I discuss below, the answer lies with a historical process shaped by both human sociocultural factors and the biology of *Bacillus anthracis*. By no means was this a one-way or inevitable process, culminating in the development of a sophisticated weapon. As we will see, anthrax probably served as a crude biological weapon as far back as a thousand years ago, and natural agricultural outbreaks still flare up occasionally around the world, even today. The historical process narrated in this book could easily have gone differently, but the anthrax we know today reflects this complex history. Moreover—and this viewpoint has often been neglected—the bacterium believed to cause the disease, *Bacillus anthracis,* has undergone major changes in its ecology and in its evolutionary pattern of development. By no means have organisms and diseases remained static while human social and cultural patterns have changed drastically over time. *Omnia mutantur.*

WHAT WE KNOW ABOUT ANTHRAX

Just as organisms and diseases change, so do our ways of understanding them. What we call "anthrax" today was once thought of as many diseases, because there were so many different signs and symptoms. *Cutaneous* anthrax, when the infection gets into the body through a break in the skin, causes the characteristic black-centered skin swelling and carries about a 20 percent untreated fatality rate. *Gastrointestinal,* or *enteric,* anthrax results from eating contaminated soil, plants, or improperly cooked meat. If not treated promptly with antibiotics, 50 percent of affected people die with severe gastrointestinal symptoms and generalized organ failure. The rarer *inhalational,* or *pneumonic,* form of anthrax causes a flu-like (but more severe) respiratory syndrome that if untreated is fatal in 90 percent of cases. The blackened skin boils often do not appear in people infected through inhalation or the intestine, making the diagnosis of the disease very confusing. By the nineteenth century, people had begun to suspect that all of these manifestations of anthrax were somehow related because the patterns of infection were the same. Victims tended to get sick when working with sick animals or certain animal products.

In animals, both livestock and wild, the most common signs are sudden death, followed by bleeding from the bodily orifices. This bleeding provides an important clue to the life cycle of *Bacillus anthracis.* Not until the 1870s could scientists watch the bacillus go through its life stages, given the earlier crude technologies of microorganism culturing and microscopes. *B. anthracis,* when living inside an animal's body (or a laboratory simulation of a body), takes on its *vegetative* form. In this form, it lives in the blood and tissues and can reproduce itself (very rapidly), resembling long threads or filaments. But expose the vegetative forms to nonbody conditions—air, coolness, sunlight—and surprised scientists peering through their microscopes would see the bacilli turn into small, round, hard *spores.* The spores had thick walls and could survive just about any rough treatment; scientists quickly found that destroying them was very difficult indeed. Only when the spores got back inside an animal's body would they dissolve their walls and once again turn into the familiar vegetative forms of the bacilli. The spore is central to the life cycle, and the primary mode of transmitting the bacillus to a new victim occurs when grazing animals eat the spores. Humans, who are not grazers, have been accidental hosts for the bacilli.

When not living in an animal's body, the bacillus survives within the protective walls of its spore, buried in or resting on the soil. Here it can stay until the next grazing animal comes along and ingests the spores. Especially after digging, flooding, or any other soil disturbance, large numbers of spores are found on the gravesite of any animal that has died of anthrax. For scientists, the discovery of the spores connected the soil of particular locations with the bacilli found in the bodies of living animals. Spores became the missing link between soil and organism, and this understanding made long-observed differences in disease processes more understandable.

In peering back at *B. anthracis'* family tree (something we have only recently been able to do), it is clear that the bacillus' soil-based life cycle relates it closely to other bacteria such as *Bacillus cereus* and *Bacillus thuringiensis.* But there is a difference: for much of our history with *B. anthracis,* it has been differentiated from its more benign cousins by its ability to kill animals and people. Somewhere in time, tens of thousands of years ago perhaps, *B. anthracis* acquired something that made it different from its closest relative, *B. cereus*—something that made it a much more efficient killer. How it did so remains a mystery, but scientists speculate that it swapped genes with another, more pathogenic microorganism to gain its lethal abilities. Once it had the ability to kill, *B. anthracis* gained a major evolutionary advantage over its relatives: it could infect a wide variety of mammalian species. Inside mammalian bodies, it reproduced itself quickly and spread over larger spaces than it would have otherwise. Of course, this could only happen if *B. anthracis* used the animal body as a reproductive chamber, escaped in a form that would enable it to survive in the harsher conditions of the outside environment, and, finally, infected a new host. This way of understanding *B. anthracis'* life cycle explains why it *must* kill its animal or human hosts. "Nothing personal," it might say, if a bacillus could talk, "I'm just making a living." But the bacillus and its life cycle form only part of the overall story. Many of the interactions between humans and the bacillus have been mediated through our dynamic understandings of anthrax, the disease.

BIOGRAPHY AND DISEASE

The genre of biography has expanded in recent years to include a variety of subjects—many of them a bit unusual. In the early twenty-first century, books in print include biographies of cities, insects, continents, languages, and even God.[1] The title of the series to which this book belongs, Biographies of Disease, hints at how this book aims to contribute something a little different to the literature on anthrax. This book focuses on scientists, physicians, farmers, wool-factory workers, and others who have experienced or worked closely with anthrax; in this sense, it is really a history of science. But *Death in a Small Package* considers anthrax, and the microorganism now believed to cause it, to be players in the history of the human-disease relationship. Thus this book narrates the history of anthrax through the lens of human actions and beliefs as they are linked to anthrax's fascinating life cycle and ecology.

There are many advantages to this approach. First, it explains much of the historical legacy of anthrax, in which the general fear of the disease seems out of proportion to the small numbers of human and animal cases. Although seemingly illogical, this fear is perfectly explainable: it has come about through humankind's historical experience with anthrax, much of which is described in this book.

Biological information about anthrax can provide new evidence for historians to consider as they analyze social and cultural events. Within the past ten years or so, computer models that simulate *Bacillus anthracis'* ecology and genetic techniques that explore its past have generated data that sometimes confirm and sometimes challenge longstanding historical narratives about disease. We must judge how best to keep various types of evidence in dialogue with one another—to "maintain an active tension . . . , using one source to interrogate, deconstruct, or illuminate the other" (in the words of Laurie A. Wilkie, a historical archaeologist).[2] For example, genetic information on *B. anthracis* used as a biological weapon can help to identify the sources of the bacteria (and perhaps the perpetrator or perpetrators of the actions, as in the case of the 2001 anthrax mail attacks in the United States). Our historical narrative is made richer by putting various sources into conversation: historical documents, maps, genetic studies, and ecological models (among many types of evidence).

A good example of this historico-biological dialogue is the spread of particular genetic strains of anthrax throughout the world in the nineteenth century. Beginning in the late 1700s, textile production became industrialized in nations such as Britain and France. As domestic livestock production failed to meet demand or as prices increased, textile mills imported more and more of the cheaper wools, horsehair, and animal hides produced in areas such as present-day Turkey, eastern Russia, and China. By the 1840s, all European textile manufacturing nations were importing large amounts of these animal products, some of which harbored the spores of *B. anthracis*. Essentially, the global trade in animal products created an ecological exchange that provided a tremendous opportunity for the bacillus. Because it was able to hitch a ride around the world in infected wool and hides, the bacillus experienced probably the greatest expansion of habitat in its history. If it were not for this expansion, anthrax would most likely be a rare and insignificant disease today.[3] At the same time, anthrax the agricultural disease assumed another identity, as an industrial disease spread by the globalization of trade in animal products. This had profound consequences for people who would otherwise never have been exposed to *B. anthracis* and who certainly did not expect to fall victim to anthrax.

ANTHRAX BECOMES A MEMBER OF THE LABORATORY AND THE HOUSEHOLD

Throughout this book, I argue that *B. anthracis* became "domesticated" to new spaces in the past few centuries: laboratories, factories, towns, and human bodies. In the pages that follow, *domestication* means bringing an organism into close relationship with human populations (often by way of animal populations). For example, along the lines of Robert Kohler's *Lords of the Fly* and Karen Rader's *Making Mice,* chapter 2 explores how the organism and its investigators adapted to each other's domestic presence in the laboratory.[4] *B. anthracis* settled into new niches, while scientists used their intimate knowledge of the bacillus' biological components to create a vaccine they hoped would defend humans against infection. Domestication brought unintended consequences, however, as the bacillus colonized new domiciles (homes and office buildings) and was sought by people determined to make it into a weapon. *B. anthracis* and the disease it causes have become part of the modern *domus,* or "home," and humans' great fear of this disease owes much to its location among human habitations and workplaces.

This use of *domestication* explicitly differs from that commonly found in other histories of diseases. In earlier studies of ecology and disease, such as William McNeill's *Plagues and Peoples,* domestication (an idea borrowed in part from 1960s ecology) meant that diseases became less and less destructive, their former extreme malignancy dampened further the longer they interacted with human populations; they became tamed. *Plagues and Peoples* suggested that the organisms that caused the disease, and the populations of their victims, had evolved to a more benign relationship such that the organisms lost their ability to kill large numbers of susceptible people.[5] Certainly some diseases have interacted with host populations in this way, but others have not. The benign evolution model never worked very well for certain diseases, including anthrax, and recently, the scientific assumption of evolution toward a benign relationship in disease systems has been challenged by biologists such as Paul Ewald. Thus, as this book concludes, anthrax will probably remain useful as a biological weapon precisely because it is unlikely to evolve toward a more benign relationship with populations of susceptible people (who are equally unlikely to evolve ways to evade *both* its lethal toxins in the near future). This redefinition of *domestication* broadens the utility of the term in light of the present scientific generation's views on disease evolution. Certainly these views, and our understanding of disease history, are open to change in the future; but for now, it is best *not* to associate domestication with the idea that evolution will direct anthrax (and some other diseases) toward a more benign relationship with humans and their animals.

Bacillus anthracis has traveled through a series of stages in the process of becoming domesticated. At each of these stages, vestiges of the previous stages continue to exist *de novo* and as building blocks of the transmuted bacillus. In other words, no matter what *Bacillus anthracis* becomes, it will always retain the material and representational traces of where it came from. First comes the bacillus as it has long existed, untouched by human hands, but ever evolving into relationships with other organisms. Like us, the bacillus must make a living, and it has responded to the selective pressures of its environment over time, such that it has never been static. Thus, when scientists began trying to bring the bacillus into human-controlled environments, such as the laboratory, a great deal of biological diversity came in with the blood and tissue samples from animals that had died of anthrax. Scientists collected these materials and in effect became bacillus farmers, seeking to grow and cultivate aspects of the complex bodily samples that provided the material basis for studying the disease.

The ability to cultivate and isolate the bacillus, name it, and characterize it took decades. Historian and philosopher Bruno Latour has characterized several stages of domestication, using the case of Louis Pasteur and his colleagues' famous development of a vaccine against anthrax. First, the Pastorians brought raw materials from sick animals (blood samples and tissues) into field laboratories and learned to cultivate and preserve the bacilli. Back home in Paris, Pasteur and his closest colleagues challenged the bacilli with heat, chemicals, and life within a series of experimental animal bodies, in order to create an attenuated, or "tamed," form of the bacillus to use as a vaccine. Then, as Latour has described, Pasteur's colleagues brought the domesticated organism back out into the field in the famous public vaccination trials at Pouilly-le-Fort in 1881.[6] By this time, cases of inhalational anthrax, then known as woolsorters' disease, had become a major topic of study for scientists and physicians in anthrax districts and industrial centers of wool and worsted production. Thus, the farm and factory joined laboratories as spaces in which *Bacillus anthracis,* due to its ability to persist for decades in any environment, had come to live. As other human spaces became homes for anthrax, even the laboratory-derived bacillus slowly escaped the control of the scientists who had nurtured and manipulated it.

ANTHRAX BECOMES A BIOLOGICAL WEAPON

To understand how *B. anthracis* became a biological weapon, we must focus on the scientists whose ideas, practices, and culture have been the conscious shapers of the bacillus over the past 150 years. This is no easy task, because much like the bacillus' ecology, the enterprise of science has varied greatly from time to time and place to place. Rather than try to detail all the scientific work done on anthrax, I discuss selected cases (some more familiar and some less) that provide windows into the changes in experimental practice and in the organism itself: Robert Koch's isolation and cultivation experiments; Louis Pasteur's vaccine trials; John Bell's investigations of anthrax outbreaks in factories; David Henderson's efforts to create an inhalational form of the spores; and the work of many investigators striving to create protective vaccines for animal and human populations.

Bitter scientific rivalries and national disputes emerged over the right to define and manipulate anthrax. Koch and Pasteur's hostility toward each other, born of a priority battle over who first established the link between *B. anthracis* and the disease of anthrax, only became magnified over time by the enlistment of scientific allies, the development of international reputations, and intensifying nationalistic rhetoric in the last quarter of the nineteenth century.[7] Reflecting profound differences in preferred experimental techniques, the two schools could not even agree on what to call the budding study of microorganisms: to the French, it was microbiology, and to the Germans, bacteriology. Scientists trained in these two traditions fanned out around the world to form the nucleus of microbiological investigations carried out in other nations with rapidly growing scientific institutions (such as the United States) and colonial institutes (the Pastorians controlled an extensive network).

After World War I, during which German-trained scientists first used *B. anthracis* as an agent of a state-sponsored biological weapons program, other nations felt compelled to develop the capability to defend against an anthrax attack (with each erroneously suspecting the others of having functional weapons). But defensive programs quickly became offensive, with *B. anthracis* at the center of military-sponsored weaponization programs because it survived delivery by airplane-carried bombs and aerosolized solutions. The intense secrecy of investigations into *B. anthracis* created many consequences, including mistrust and loss of accountability. But, as historian Brian Balmer has shown, secrecy has also expanded the realm of scientific experiment.[8] Fishermen straying accidentally into an aerosolized weapons testing site off the British shore, the pustule on a laboratory worker's neck, and the experimental vaccine injected into a wool-factory worker's arm were all events that scientists' secrecy converted into experiments. *B. anthracis* experienced this conversion of life into science as an expansion of its habitat and its genetic diversity. While few of the scientists who had enabled this expansion recognized it, the ecological and ideological consequences for human–*Bacillus anthracis* interactions have been profound. *Bacillus anthracis,* in the process of continuing to make a living, has become a denizen of human-controlled environments that has nonetheless escaped our control.

❋ ❋ ❋

This book, being a short history of a complex topic, does not strive to be an encyclopedia of information about anthrax. Indeed, when considering events of the past century alone, this would be impossible. Due to the top-secret nature of biological weapons programs, most historians have minimal access to the oral histories, interviews, and original documents that would allow us to create comprehensive and authoritative analyses about the development of biological weapons. It can even be difficult to identify the scientists involved, since those working in covert operations could not participate freely in scientific exchange. They did not discuss their work with their peers, nor did they publish their results openly. If the majority of their careers were devoted to working in secret, then their histories usually remain difficult to recover. Many scientists who have worked in military programs with *B. anthracis* are understandably reluctant to discuss their work.

Nonetheless, *Death in a Small Package* organizes the many histories of *B. anthracis* and anthrax, the disease, into an argument based on the available historical record—an argument that impresses on us the importance of understanding historical processes in creating living weapons and the resultant diseases. These historical processes begin with ancient stories about anthrax, a disease that has long cursed people's relationships with their domesticated animals.

CHAPTER ONE

Infectivity and Fear

Charbon *and the Cursed Fields*

❋ ❋ ❋

An animal which has a strong charbon infection, [temperature of] say 105 or 106, lives or dies within 3 days. [The place where it dies] will remain for years a spot of troubles.

Diary of Louis Leonpacher [July 14, 1921]

What exactly *is* anthrax? A disease, to be sure; but what we now call anthrax has had many names and even different symptoms in different sufferers. People have thought about and reacted to anthrax in many ways over time, creating changing cultural definitions of the disease. The biology of the disease and its causative agent have also probably changed over time. I explore this question of what anthrax is in this chapter, ranging widely across time periods and geographic locations to provide snapshots from anthrax's long history.

Two particular characteristics show up again and again, and they are suggested by the quotation above from Louis Leonpacher's diary (Leonpacher, a Louisiana veterinarian, calls anthrax by its French name, *charbon*). First, anthrax has long been associated with infections in grazing animals, and second, anthrax persists in particular places for a long time (what Leonpacher called a "spot of troubles"). The first characteristic may seem obvious: throughout much of the past two thousand years, anthrax has been primarily an agricultural disease that affected livestock and their caretakers. Mysterious outbreaks of any disease in humans usually were explained and named according to the sufferers' symptoms, but the symptoms of anthrax could vary. Thus, a concurrent outbreak in animals helped to identify the problem as anthrax. Although it readily infected people, a disease could only be "anthrax" if animals were also ill and dying.

The second characteristic that defined anthrax lay with its localizability to certain places. Anthrax was a *telluric* disease, meaning a disease of the soil. One outbreak of anthrax would contaminate and pollute a pasture for years to come; yet the disease was fickle. Anthrax could sicken animals every year or remain dormant for a while and then make a sudden appearance. In France through the medieval and early modern periods, the affected pastures were known as the *champs maudits,* the "cursed fields." Animals grazing the champs maudits could be thriving one day, then suffering and mysteriously dead the next. The curse followed the animal's body to the humans who skinned it and cut up and ate the meat. Anthrax struck and destroyed without warning, from some seemingly invisible cause, almost inescapable if one lived on or near the champs maudits. It turned the blood of its victims dark and caused blackened skin pustules (after the fourteenth century, this symptom struck terror into Europeans in part because it resembled lesions of the Black Death).

The centuries of human experience with anthrax fostered its fearsome reputation as a merciless killer. Its arrival terrorized the human community, whose members attempted to meliorate its effects by removing animals from cursed pastures, praying, and using various medical and surgical treatments on the sick. This deeply enculturated fear would emerge again in the twentieth century, helping anthrax to become a potent weapon of terror. Through much of human history, anthrax-like diseases threatened people's livelihoods and their very lives, motivating them to try to explain and control the disease. Scientists in the twenty-first century have been no exception, as continuing interest in anthrax has stimulated them to study the genetic footprints of the disease.

Scientific studies give us a source of evidence (along with historical documents) to explore the origins and spread of the disease. Since the 1990s, scientists have been increasingly able to distinguish between different strains of anthrax and to trace the past movements of the disease. They have been able to relate anthrax found in Europe with that found in South America, for example, and to estimate when the disease traveled from one place to another. Thus genetic evidence provides information that we can combine with historical documents to piece together the history of a disease, and I have used it in this chapter to augment other historical sources.

But before we can understand the history of anthrax, it is important to think carefully about how people named and defined the disease over time. The historical processes of naming and differentiating the various manifestations of anthrax encompassed a wide range of observations and beliefs. Analyzing the language that people used to identify and characterize disease provides a window into how they understood the workings of disease. Where did it come from, and why did some people and animals sicken while others remained healthy? What could be done to respond to a disease outbreak or to prevent one in the first place? The answers to these very practical questions depended on one's basic assumptions about the disease. Anthrax was a disease of soil, animals, and people, and its complex ecology has powerfully influenced its ever-changing identity since ancient times.

CHOOSING NAMES FOR ANTHRAX

What we think of today as anthrax did not take on its current identity as a defined and recognized disease syndrome until around 1800. Therefore, when historical commentators write about "anthrax," we cannot be sure they are describing the same disease that we now call anthrax. We can only approximate, based on the behavior of the disease that they described. From ancient times, surviving texts, language etymologies, illustrations, and husbandry practices demonstrated a deep knowledge of disease and how to respond to it. For example, anthrax caused different sets of symptoms according to how the disease was acquired. Early modern classification systems of disease differed greatly from our system today. Diseases were often classified according to the signs and symptoms they produced, and anthrax-like diseases often had particularly colorful and gory names. "Malignant pustule," a case of anthrax that began with a skin boil, fit within the category of what we would call dermatology, a skin problem. In the French language, for example, *clou charbonneux* (*le clou* meant carbuncle; literally, "nail") described a dark-colored skin pustule, or carbuncle, that may have been due to infection with the microorganisms that we associate with anthrax or gangrenous bacteria, or to trauma from a spider or scorpion bite.

A case of anthrax that began with a fever, however, often quickly killing the animal, got classified differently. "Splenic fever" and other names for anthrax-like diseases reflected an understanding of the whole-body disease as being an essentially different condition from the blackened ulcers. This meant that the feverish, whole-body form of anthrax (in French, *fievre charbonneuse*) was seen as more closely related to various bacterial septic infections, viral pneumonias, and numerous other whole-body problems than it was to malignant pustule, the skin form of anthrax. Why should these historical distinctions matter? Practically speaking, before the late 1800s, the treatment for a condition depended not so much on the theory of causation (the category we often use today), but on the prescribed treatments for conditions in that particular category. For example, anthrax that manifested itself as a dark-colored carbuncle got treated locally with poultices, lancing, and cauterizing, but anthrax that began as a fever likely required bleeding and blistering the patient instead. Thus, "anthrax" was not a constellation of signs and symptoms related to each other by the presence of a bacillus, as it is today; instead, there were many anthraxes that did not seem logically related to one another. Each form of anthrax looked different and was treated differently. This was not a problem for most people, who had no reason to think these various conditions *ought* to be related to each other.

Consequently, the names that people around the world have used for anthrax tell us much about their perceptions of the disease. An anthrax-like disease outbreak would often begin in populations of domesticated herbivores, such as cattle, goats, sheep, and horses. Ibn al-Awamm, the twelfth-century Moorish author of a well-known agricultural treatise, thus designated anthrax *Dā el-Bakar,* or "cow sickness." The Masai differentiated between anthrax in cattle, *Em bujangat,* and anthrax in goats and sheep, *Eng ea nairogua.* People exposed to infected animals in an agricultural setting often suffered from the malignant carbuncles, and the disease was named accordingly. In Siberia, with large populations of infected horses, anthrax-like disease in people was called *yasva* (loosely translated as ulcer or sore).[1] Names in the classical Western tradition also described the signs of skin disease: the Greek *anthrakos,* meaning "coal" or "carbuncle," and the Latin *anthrax* for "carbuncles" or "malignant boils." Names for the disease could differ depending on how many victims it affected. In medieval and early modern Europe, outbreaks of anthrax-like diseases in groups of animals and people were called "black bane," with "malignant pustule" being used to describe individual cases of the pustules.

The French name used by the Louisiana veterinarian, *charbon,* translated to mean "carbon" or "charcoal," and it likely referred to the blackened skin lesions and darkly colored blood of anthrax-affected animals. In medieval and early modern France, this was laypeople's name for the outbreaks of pustules in livestock (and the name has continued to be used today for all types of anthrax).[2] Names for the more deadly whole-body forms of anthrax in many cultures reflected the high fevers and sudden deaths in affected animals: the Arabic *Atshac, al Humrah;* the French *fievre charbonneuse;* and the Latin *sacer ignis* (sacred fire). In some places, names for anthrax also reflected the idea that it was a transmissible type of poison. The Dutch *venijn* and the Friesian *fenyn,* for example, translate into English to mean "poison."[3] Prior to the mid-1800s, these names reflected general ideas that diseases resulted from the pollution or poisoning of the body by unknown agents. This theory would have made especially good sense in the case of anthrax, which animals seemed to acquire by grazing on polluted pastures.

Many cultures' names for anthrax referred to a specific organ—
the spleen. This probably reflected the fact that the spleen is the
internal organ most drastically affected by whole-body anthrax.
(During one postmortem several years ago, I had the unnerving
experience of picking up the spleen and having what should have
been a solid organ literally drip through my gloved fingers.) I have
used one of these terms already: *splenic fever,* to refer to the whole-
body infection. In Constantinople, a location of repeated outbreaks over the centuries, the ancient name for anthrax (both the
skin and whole-body forms) was *dallack* (according to British
physicians stationed there in the nineteenth century). Dallack existed in several species of animals and infected humans who tended
those animals or processed animal products. The disease was so
common during the Victorian period that a branch of Turkish
medical practice specialized in it—the *dallackjis,* or anthrax doctors. The word *dallack* meant "spleen," and the disease was so
named "because that is the organ which is diseased in the animal
from which the infection comes." Thus the dallackjis recognized
the pathology in the animal's body (perhaps by conducting postmortem examinations) and connected it with the skin disease in
humans.[4] Late nineteenth-century South African settlers reported
the presence of the disease there and translated its local name to
milksiekte, or "spleen illness," in Afrikaans.[5] Another common
Dutch name for anthrax was *miltvuur* and, in German, *Milzbrand,* both translating to "spleen fire" or inflammatory death of
the spleen. In the early nineteenth century, the English designation "splenic fever" referred to both a common clinical sign during life (the fever) and the organ that itself died (the spleen), thus
killing the animal.

This list of names for anthrax reminds us that people understood quite a bit about the disease even before the scientific developments of the mid-1800s. They recognized that it affected both
domestic animals and humans, and there were many theories
about how this could happen (handling sick animals or breathing
the same bad air, for example). Anthrax's continuing presence in
the soil of the champs maudits is perhaps the most important

feature of the disease's ecology, and the common practice of moving animals to unpolluted pastures would have been an effective way of breaking the disease's cycle of infection. Also, several names characterized anthrax as a disease that destroyed the spleen, implying that early observers had studied the internal anatomy of those who had died of it. Several physicians, philosophers, and natural historians had a particular interest in anthrax, and their surviving writings help us to understand how our present-day concept of anthrax developed.

UNIFYING ANTHRAX

In the waning years of the *ancien régime*, French physicians had good reason to be interested in anthrax. During this period, the mid to late 1700s, some French workers in the nation's busy textile mills contracted charbon, either the pustular or whole-body form, or both. A Burgundian physician known by the name of Jean (some authors call him Nicolas) Fournier puzzled over the pattern of illnesses suffered by these workers. Fournier began to view anthrax, and thus disease classification overall, quite differently. As a young physician in the 1750s, he had experienced outbreaks of *peste de Marseille* (bubonic plague) and charbon. During the latter outbreak, Fournier consulted with his older colleague Dr. Verny, who seems to have been a major influence on his thinking about anthrax-like diseases. Verny had noticed that not all skin pustules and carbuncles behaved in the same way, and he theorized that they had different causes or interactions with the body. Verny had begun to distinguish among different types of darkened swellings on the skin—*charbon malin,* clou charbonneux, and *charbon simple*—which likely had different causes because the patterns of disease differed (and were probably caused by different bacteria, as we now suspect). In the villages around Montpellier, Fournier observed several cases of whole-body anthrax caused by the ingestion of contaminated meat and began to think of them as related to the concurrently appearing skin forms of charbon. Some people with the characteristic skin lesions developed the whole-body disease, he noticed. All had either eaten the meat of animals that died of the disease, handled such animals, or worked in the factories with animal wool and hair. Fournier speculated that the disease could manifest itself in different ways, depending on its nature and that of its victim.

Indeed, Fournier was envisioning a new classification for anthrax, one that broadened the criteria required to determine the category of disease.[6] Not only signs and symptoms, but also the disease's process of infection, determined its classification. The classification framework included both human and animal disease, and groups as well as individuals. For most physicians, the individual patient suffering from a terrible disease defined the condition, but for Fournier, anthrax was a much more extensive phenomenon—something that we would call an ecological system of disease encompassing many animal species. As epidemiologist David Morens has characterized Fournier's way of thinking, "[Fournier believed] that he was observing a spectrum of conditions larger than just cutaneous [skin] *charbon malin.*"[7] That spectrum included disease that appeared at certain times of the year, affected groups of grazing animals, caused some combination of fever and skin swellings, and could then produce the condition of malignant pustule, or fievre charbonneuse, in the humans who handled the animals or ate the meat from those that died.

Fournier's thinking brought together several previously disparate conditions under one category of classification and made charbon what we would now call a "specific" disease. Fournier focused on two divisions of anthrax-like diseases defined according to his idea of causation: spontaneously appearing disease versus contagious charbon. His categories were also differentiated by severity and class status. According to Fournier, the spontaneous form (*charbon spontané*) was a skin disease of peasants who lived and worked in unhealthy conditions. It was not contagious per se. Instead, it manifested itself when the conditions were fortuitous: filth, poor nutrition, and working in the soil under the hot sun. Contagious charbon (*charbon contagieux),* on the other hand, appeared only in people who had handled animals ill with charbon, or their wool, hides, or meat. Contagious charbon seemed more serious to Fournier, because it could result in death, and its infectious particles (*atomes de ferment charbonneux*) could travel through space and time on contaminated animal products. Contagious charbon would clearly be a hazard difficult to avoid even in the upper sectors of a meat-eating and wool-wearing society. Fournier condemned "mercenary" stock raisers and butchers who sold at decreased prices the meat of animals who had died of charbon, thus infecting poor people who could not afford better.[8]

In 1769 Fournier published a booklet about his theory on how various diseases should be grouped together in the category of charbon. It apparently did not make much of a splash, since it appeared at a time when advocates of many disease-causation theories competed with each other; Fournier's was only one of many theories. Moreover, the system of categorizing diseases by signs and symptoms, not by etiology and epidemiology, was well ensconced. Why abandon the certainty of classification by symptoms for Fournier's complex and uncertain classification system? However, Fournier's theory had its advantages. With its emphasis on the epidemiology and infectious process of contagious charbon, it would have made sense to physicians and others who had experience with anthrax. The idea that charbon was contagious between animals and people was underscored the next year (1770) by a devastating epidemic of animal and human anthrax in the French colony of Saint Domingue. Fournier had linked disparate illnesses to create a specific disease grouping, and he had intimated that fermentive particles could transmit it between animals and humans—both radical ideas at the time but very helpful to anthrax investigators who would follow Fournier.

The idea of a new classification for anthrax began to catch on in France. Fournier's interest in anthrax did not wane, and he was an important force behind a 1780 essay contest on malignant pustule in animals and humans sponsored by the Academy of Sciences of Dijon. Several of the essays continued along Fournier's track of trying to distinguish true anthrax *pustules malignes* from other, less deadly pustular diseases.[9] About a decade later, Philibert Chabert, the director of the French veterinary school at Alfort, published another important treatise that sought to relate different clinical signs and symptoms of livestock into one syndrome. He described the various anthrax syndromes in different species.[10] Chabert categorized the *fievre charbonneuse* with *carbuncles symptomatiques,* the blackened swellings, to distinguish them from more benign and nonanthracoid carbuncles (although this classification would later be revised after *le charbon symptomatique* was attributed to clostridial bacteria, not to the *Bacillus anthracis*).[11]

Fournier, Chabert, and others had figured out that the differing sets of symptoms *were* all anthrax and thus had a created a new way of understanding anthrax as a unified disease (figure 1.1). However, confusion persisted over what was not anthrax. Take, for example, the mystery of "anthrax of the tongue," also known as *glossanthrax, charbon de langue, Zungenbrand,* and *tongblaar.* Glossanthrax was an aggressive and fatal disease that affected cattle, horses, pigs, and chickens, and it caused terribly grotesque symptoms. Blisters on the affected animal's tongue turned black, burst, and spread gangrene to other parts of the tongue, which often fell out. Affected animals then died within a day. British veterinarian George Fleming recorded a vivid picture of the horrors of glossanthrax: "the fields were covered with tongues."[12] Outbreaks in Europe were recorded in the 1680s, 1730s, 1780s, and early 1800s, but the twenty-first century veterinary literature does not describe this disease, at least not as a type of anthrax.[13] French veterinary epidemiologist Jean Blancou has suggested that glossanthrax may have been a form of true anthrax that disappeared 150 years ago, although he cautions that early clinical and disease descriptions are confusing.[14]

In retrospect, glossanthrax may have been attributable just as easily to what we would call a virus with a secondary *Fusobacterium necrophorum* infection, or it may have been foot and mouth disease.[15] Most likely, our twenty-first-century classification of glossanthrax would differ remarkably from that of early nineteenth-century authorities. Chabert, Henri Delafond, and other French authorities believed glossanthrax to be one of the true anthrax diseases; so did German and Dutch authorities on animal diseases

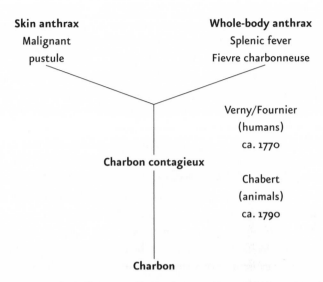

Figure 1.1. French classification system for types of anthrax. The definition of the disease *anthrax* (or *charbon*) familiar to Westerners today is a conglomeration of what were formerly understood to be many different diseases. In the late 1700s, French physicians and veterinarians asserted that skin anthrax and whole-body anthrax were related to each other and somehow contagious (a radical new way of defining the disease).

(Alexander Numan and F. C. Hekmijer, for example). By Delafond's time, around 1840, the disease seemed to have disappeared. The last case labeled glossanthrax that I found in the veterinary literature occurred in a polar bear in 1911 at the London Zoo, as reported in the British journal *Veterinary Record*.[16]

At this point, you might be asking, "so what *was* anthrax, then?" The answer, of course, depends on the period being considered. Fournier and Chabert's published works helped to remake the identity of anthrax in the late 1700s. The continuing confusion about disease classification is thought to have been resolved in the mid-1800s by the "discovery" of the *Bacillus anthracis* (the topic of the next chapter). Associating this bacillus with anthrax again introduced a new way of classifying the many diseases of animals and humans that looked like some form of anthrax. However, the symptoms and characteristics of the disease's spread have remained important identifiers. Disease identity depends on many factors, and in the case of anthrax, identity could not be easily reduced to a strict bacteriological etiology model even after the *Bacillus anthracis* was identified in the 1870s. One ancient factor associated with the disease, its persistence in certain geographic places, has long been an important identifier of anthrax.

ANTHRAX DISTRICTS: THE CURSED FIELDS

It was fairly common for diseases to be associated with certain climates, or certain geographic locations, and anthrax had special features in both categories. Anthrax emerged from *champs maudits,* or cursed fields (which could be cursed by a poor location or climate).[17] For centuries, livestock raisers familiar with the disease's somewhat erratic appearances have associated it with changing climatic conditions. Spring floods, dry summers, rainy autumns, and cold spells all could be indicators of a bad anthrax year, depending on one's experience and point of view. The one thing everyone could agree on was that somehow anthrax-like disease persisted in certain fields, pastures, and regions—areas that became known in the early modern Western world as anthrax districts.

The first of these specific places lay in the Fertile Crescent of Asia Minor, the cradle of animal domestication between 5,000 and 10,000 years ago. As historians of the Near East have defined it, the Fertile Crescent began at the mouth of the Tigris and Euphrates rivers in the Persian Gulf, stretched north to Lake Van in present-day Turkey, and then curved to follow the shore of the Mediterranean Sea, connecting with the mouth of the Nile in Egypt. Plants, animals, and humans have long thrived in this region of abundant water and rich soil. Based on recently elaborated genetic analyses, the biological origins of what we now call *Bacillus anthracis* are thought to precede the Holocene period 10,000 years ago (a long time in human history, but relatively recent in evolutionary terms).[18] Anthrax-like diseases probably first established themselves as an important force in human history in the Fertile

Crescent for two reasons. First, a critical mass of susceptible animals is required for outbreaks to occur. Early attempts at domestication created unusual concentrations of these susceptible animals. Once malignant pustule or whole-body anthrax was established in a place that harbored dense populations of animals, it persisted at that place for decades and even centuries, periodically reappearing in new populations of animals and people.

Second, these disease outbreaks mattered because of the importance of domesticated animals to Fertile Crescent societies. Human populations accumulated wealth with their animals, and centers of commerce in the ancient world depended on animals for transport as well as food. Outbreaks of disease meant economic and personal hardship, occasionally to the extent of starvation, as well as illness and death from the disease itself. Animal disease was discussed, recorded, and interpreted through a variety of human discourses. Although we can never be certain of the identity of ancient diseases according to our present definitions, the epidemiological patterns of what we think of as anthrax were described by natural historians and physicians in writings that survive today. Their observations included animal outbreaks characterized by acute deaths, skin boils, blackened blood, and the sickening of people associated with those animals.

Interactions among culture, climate, and geographic location in disease outbreaks have also mattered to historians seeking to identify ancient diseases. In the most commonly cited early example of an anthrax-like disease, the fifth biblical plague, the affected cattle were grazing on rich lowlands, whereas cattle that were grazing on the poorer but higher soils did not manifest the disease. Lower pastures subject to flooding, as these purportedly were, have long been thought more prone to anthrax. (Current scientific knowledge suggests that floodwaters brought the spores of the infectious organism to the soil surface.) That the cattle grazing on the poorer, higher pastures belonged to the downcast Israelites (with the richer lowland pastures reserved for the ruling Egyptians' cattle) leads to another interpretation of the same events as a plague visited upon the Egyptians by the Israelites' god (who sought to avenge Israelite enslavement). That the plague appeared only in certain places and not others made it a telluric event that had both ecological and cultural significance.[19]

As crops and domesticated animals spread outward from the Fertile Crescent, so did diseases. By 500 BCE, the animal healing literature of India focused on diseases of cattle, elephants, and other domesticated animals. Historians believe one of the diseases to have been anthrax, due to the description of its carbuncular nature, its transmission to humans, and its location in certain places.[20] From Egypt, sick animals probably transported anthrax-like diseases down the Nile and along trading routes to various areas on the continent of Africa. Animal bodies and soils from Mesopotamia also disseminated these diseases northward and around the Mediterranean world, where they appeared in the literature of the Greeks and Romans and from which they probably entered Europe. Virgil's *Georgics* (3.440–566) famously recorded a series of animal plagues, including a very detailed description of an anthrax-like disease in horses, cattle, and sheep that also affected humans who wore contaminated fleeces (563). The disease apparently came down from the Alps and into Italy, where it devastated animal populations.[21] Virgil's use of the terms "black blood" and "sacred fire" (*sacer ignis,* a term that has been used for other diseases as well) has persisted for two millennia in the names of anthrax in several languages. One of these outbreaks of anthrax-like disease, circa 500 BCE, purportedly destroyed almost half the human and animal populations of Rome. The disease remained a problem in the same geographic areas for centuries.[22]

The Mediterranean area, comprising twenty-five contemporary nations on three continents, proved to be an enduring location for anthrax and other animal-human diseases due to the concentrations of human and animal populations, intensive trade, agricultural practices, wars and particular political events, and the parameters of soil and geographic diversity.[23] Political events influenced the spread and interpretation of disease as well: outbreaks in areas outside the Fertile Crescent were noted by Greek natural philosophers and physicians who lived or were serving abroad.

They interpreted these diseases according to what they knew, so animal diseases that looked and acted like what was known as anthrax in Greece was labeled anthrax. Apsyrtos, chief military veterinarian in the army of Constantine the Great, recorded outbreaks of fatal pustular diseases among cattle in the Byzantine period and recognized this as anthrax.[24] A century later, Publius Vegetius described anthrax in cattle as a specific disease in the *Digesta Artis Mulomedicina*.[25] Muslim trade routes, traveled by caravans of animals used for food and transport, connected northern Africa and southern Spain with the Middle Eastern and central Asian anthrax districts. Had anthrax traveled with successive demographic shifts between the Middle East, Asia, and Europe? Or did something like it already exist well beyond the Fertile Crescent millennia ago, to become "anthrax" through the interpretations of long-dead observers? We cannot know for certain, but one thing is clear: by about 1000 CE, anthrax-like diseases had become well established throughout the Old World.

Later outbreaks of anthrax-like diseases tended to follow the patterns of Roman settlement. What is now present-day Germany suffered from a major epidemic in 943 CE; in the British Isles, the infection had traveled to Ireland by 1224.[26] By the end of that century, German, Italian, French, and British sources all described disease outbreaks that resembled anthrax.[27] Veterinarian George Fleming, writing in the late 1800s, recorded large *epizootics* (outbreaks in animals) of "anthrax fever" in England in 1252 and 1319. Both outbreaks caused great mortality among people, cattle, and the dogs and birds of prey that fed on the cattle carcasses. Contemporary observers called the disease "blain" or "black-blain." Fleming's book reflects the causation theories and interests of his time: it correlates climatic and geological events with disease outbreaks. Fleming wrote that the 1252 and 1319 epizootics followed peculiar climatic conditions and were believed to have arisen from infected pastures.[28]

These terrible outbreaks among animals wrought havoc on the human population as well. A 1613 outbreak in southern Europe, for example, killed an estimated sixty thousand people.[29] The epizootics and epidemics seemed to intensify in the 1700s, occurring not only in older anthrax districts but also in waves, moving from place to place with imported animals. Even though anthrax-like diseases had long been established in western Europe, the temptation was great to blame outbreaks on "foreign" populations of animals and people. In the early 1710s, anthrax-like disease decimated the cattle population of Hungary. When it hit France in 1712, French peasants angrily asserted that the charbon had come from Hungary (despite their long experience with it on their native soil).[30] Reflecting on the overall pattern of anthrax outbreaks, François-Hilaire Gilbert, a French professor of agriculture and veterinary science, wrote in 1795 that the disease appeared to be more powerful and seemed to affect more areas than it had in the past.[31] Europe suffered these severe anthrax outbreaks during the busiest years of trade with its colonies. That coincidence, if it was one, altered the course of colonial relations in areas of the New World.

ANTHRAX EXPANDING

We do not know if anthrax-like diseases existed in the New World prior to European contact, although I think European livestock most likely brought these diseases with them during the years of colonial activity (about 1500 to 1800). There is some evidence (both scientific and historical), however, to support an earlier importation of the *Bacillus anthracis*. Recent genetic studies of anthrax in North America and Russia have found close similarities between the two strains of bacillus. These findings suggest that anthrax might have migrated with people and animals across the land bridge during the Holocene period and become established in North America.[32] Populations of deer and other wild animals may have provided just enough host-species density to maintain low levels of endemic infection. The anthrax-like diseases may have been able to spread southward because South Americans maintained domesticated populations of susceptible animals, alpacas and vicunas. But then, why would the disease outbreaks in livestock have become so much more devastating after about 1500?

The numbers and density of animals may have increased enough in colonial times to activate already existing anthrax, or we may simply lack historical and scientific evidence of earlier outbreaks. This theory remains only conjecture, however, and other data point strongly to a relatively recent introduction of anthrax-like diseases from the Old World to the New.[33]

Based on what we know about the ecology of anthrax, specifically its need for large settled populations of herbivorous animals, South, Central, and North America were probably spared the destructive outbreaks so common in Asia and Europe before 1500. Several pieces of evidence point to the European colonial introduction of anthrax to the Western Hemisphere after 1500. Native Americans, immigrant Africans, and Europeans all referred to anthrax as a "new" disease in the Americas. Only after the Spaniards brought cattle and pigs to the Caribbean in the 1500s did anthrax became a major threat to animals and people in the New World. Along with sick animals came contaminated meat, wool, and hides, and potentially contaminated soil traveled in the ships' ballasts and cargoes. Anthrax-like outbreaks from the Old World first appeared on Caribbean islands, such as Santo Domingo, that were important stopping-off places for Spanish and French ships bound for the mainlands of the New World. Although some historians have suggested that anthrax arrived on the North American Gulf Coast with ships that imported it from the valley of the Nile, this route was rarely used.[34] Instead, people and animals most frequently traveled between the islands and the mainland, and anthrax most likely traveled back and forth between these locations as well.

Anthrax was a disease in motion in the New World, centered on islands that witnessed dramatic human and animal demographic disruptions and that served as mixing vessels for diseases. Santo Domingo (also known as Saint Domingue and, at present, Haiti and the Dominican Republic) provides an important example. Santo Domingo may well have been the entry point of what the European colonists called charbon to the Western Hemisphere in the modern period. European domesticated animals arrived in the Caribbean with Columbus in 1492 and 1493, and Spaniards established settlements and eventually *hatos,* or cattle ranches, on Santo Domingo. By 1501, imported African slaves joined the Spaniards and the native Arawak-speaking Taino population. Despite the devastation of smallpox and Spanish depredations, the Taino fought to preserve their vibrant culture and well-established villages. Depending on the observer, sixteenth-century Santo Domingo must have been an exciting and terrible place: the continual and brutal turmoil among the local Spanish, African, and native populations periodically interrupted by the regular plundering visitations of Dutch, British, and French pirates. By the end of the seventeenth century, the native Taino population was extinct, plantation owners and ranchers had brutally suppressed periodic African insurrections, and the western part of the island was ceded by Spain to France (with a consequent name change to Saint Domingue). All these events took place against a backdrop of disease outbreaks in both people and animals.[35]

Perhaps the most devastating outbreak of what we assume to be anthrax occurred in Saint Domingue in the 1770s. Writing in 1826, historian Michel-Placide Justin described an epidemic of charbon that killed an estimated 15,000 people and uncounted thousands of cattle and other animals on the island in 1770.[36] By this time, the human population included various combinations of African freedpeople and slaves, Spanish merchants and ranchers, and French colonists and planters. Animal populations included the cattle on the hatos and large herds of feral animals (cattle, pigs, and goats) that escaped the ranchers' control. The feral animals provided a living for poor people who caught and slaughtered them for meat to sell in the settlements and on the plantations. These people also tanned hides and processed tallow for shipment to Europe, a major source of income for islanders who did not own plantations or ranches.[37] Periodically, this important economic system got disrupted by disease in the animals.

In June 1770, amid an outbreak of an anthrax-like disease in both the domesticated and the feral cattle, Saint Domingue experienced a deadly earthquake. This disaster led to social chaos and famine, and starving people soon began eating meat taken from wild and domesticated cattle that had died of the disease. Justin reported that the meat was consumed in the form of *taussau,* salted or smoked and usually uncooked, and that the African peoples avoided eating taussau unless they had nothing else. The disease spread quickly in Africans who had purchased and eaten taussau, and charbon appeared in the neighboring dwellings of the Spaniards and along their travel and trade routes. Justin wrote that the infected meats spread the *germe* of charbon to the human population, creating an epidemic of the disease. Justin (or his unknown sources) may have exaggerated the death toll, which he placed at 15,000; but the path of the disease seemed clear to him. As epidemiologist David Morens concluded, Justin's report "linked the distribution of contaminated beef to the geographic spread of human disease . . . consistent with intestinal anthrax, a disease associated with high mortality."[38] Further outbreaks of the terrible disease occurred throughout the 1770s and into the 1780s. This chain reaction of chaos, with charbon as its igniter, may have contributed to the social turmoil that led to bloody revolution and the dissolution of the colonial government in the early 1790s.[39]

The charbon outbreak of 1770 was probably not the first and definitely not the last. Outbreaks occurred almost every year through the 1770s (recorded by the French colonial authorities and Saint Domingue's Cercle des Philadelphes, of which Benjamin Franklin was a member).[40] Large areas in Saint Domingue had become anthrax districts, and from here anthrax spread through trade routes to other areas. Henri-L.J.B. Bertin recorded an epizootic in Guadeloupe in 1774; cattle, mules, and people in Grenada died in large numbers in 1783.[41] Looking back a century later, British veterinarian George Fleming quoted a Mr. Chisholm, a medic in Grenada in the 1780s: "The flesh of the cattle that died, being dug up and eaten by the negroes, proved most dreadfully septic, producing a pestilential carbuncle, attended by a malignant fever." Fleming obtained another account of the disease from the mid-eighteenth century: "Among the distempers which infect the horned cattle, there is one of a very contagious and pestilential kind, for a beast shall seemingly, by his feeding heartily, and in appearance, be otherwise well, yet in a few hours, without any symptom of a previous disease, drop down and die. These when dead, are by the most judicious planters immediately buried, and often there is a watchman appointed to prevent the new-bought negroes and others of the poorer sort from digging up the carcases and feeding upon them; for when this happens it generally costs their lives."[42]

One contemporary observer believed the disease to have come to Saint Domingue with horses from the North American mainland, but the transmission pattern most likely was reversed because susceptible populations of animals occupied the islands before the mainland.[43] Recent scientific studies and genetic investigations reinforce this link. Scientists have found that a particular strain of *Bacillus anthracis* found in the Caribbean and in Central America is also the major strain found in the United States and Canada. These data support the idea that the bacilli traveled between these areas at the time of colonization and that a distinctive New World strain has been preserved for the past four hundred years.[44] Although the data do not indicate the direction in which the disease traveled, here is where the historical documentation supplies information that the scientific methodology cannot. Historical records show that the necessary large populations of animals existed first on Santo Domingo and other islands, which became anthrax districts and most likely were the source of the anthrax strain disseminated to North and South America.

Historical evidence also indicates that mainland populations had not seen what the colonizers called anthrax or charbon before it traveled with European ships. After the tremendous Caribbean outbreaks of the late 1700s, the disease moved inward on the mainland of North America through the early 1800s. Although historians need to do more research on this point, anthrax bacilli likely traveled inland along the routes frequented by animals, such as cattle trails and up the major rivers. Even the descendants of the European colonizers had little familiarity with anthrax-like diseases. Reports of outbreaks first appeared in the medical literature of North America in the early 1800s (there were no veterinary journals at that time). In 1824, for example, physician John Kercheval of Bardstown, Kentucky (then a relative wilderness), described "an anomalous and extraordinary disease" that appeared among cattle, sheep, and horses (some of which had come from the deep South). Kercheval described the "swellings" on affected animals, the blood that was so "dissolved that it transuded through the pores of the skin," and the sudden deaths.[45]

The disease could affect humans, too: they suffered from a "derangement of the entire system" that followed the development of a "small and circumscribed vesicle" whose center became "livid and black." Kercheval reported that "no one was affected with it, who had not been previously engaged in flaying or otherwise handling and touching the carcase of an animal that had died of the distemper described." Kercheval heard no reports of African American victims, which he attributed to "the morbific virus not coming in contact with . . . a scratch, or abrasion of the skin." (More likely, African Americans received little or no attention from Kercheval or his physician colleagues and cases in this population went unnoticed.) Kercheval concluded, "Alike novel in its character and unique in fatality, it is viewed here, as a new disease."[46] The eminent Philadelphia physician James Mease carefully noted Kercheval's description in 1826 and warned of the dangers of flaying cattle that had died of "malignant" diseases such as the new disease described by Kercheval.[47]

Indeed, anthrax was a "new" disease because it was situated in a new sociocultural and ecological context. Kercheval's account described the loss of animals, people's fear of developing the illness, and the disquieting observation that the disease mainly affected Kentucky's white citizens (always fearful of being outnumbered by the largely enslaved African American population). In the New World, anthrax followed not only cattle culture, but also slave culture. The intricate economies of the colonies intertwined populations of people, animals, and disease that traveled great distances ecologically as well as geographically. Kentucky had become part of larger biological and economic systems by 1824.

Linked by rivers and gradually improving inland routes, Kentucky exchanged soils, plants, animals, and people with Louisiana, the Caribbean, and the Chesapeake. The rapid growth of the state's population and its animal economy created uniquely propitious conditions for populations of *Bacillus anthracis* to expand. Once brought to a place, the bacillus could then establish itself in the soil and begin its cycle of persistent recurrent outbreaks—it had become domesticated to a new location.

ANTHRAX WEATHER AND ANTHRAX SOIL

What determined the pattern of anthrax outbreaks? Why did the disease wax and wane instead of wreaking havoc every year? In part to try to answer this question, ancient observers associated anthrax's identity as a telluric disease with the climatic cycles that affected those soils. The biblical story of the Egyptian plague had stressed the floods that enriched (and probably infected) the cattle pastures; Sanskrit texts describing an anthrax-like disease depicted it as peaking in the dry and hot summer. Cycles of drought and flood appeared again and again in accounts of bad anthrax years, especially in Europe.

For example, George Fleming blamed a major European epidemic in 1756 on the climatic conditions. The winter had been mild, and the spring and early summer very dry and hot. Heavy storms hit in June, flooding many areas. Then, as Fleming recounted, "the cattle sickened suddenly in the pastures . . . and died in from twenty-four to thirty-six hours. The game, horses, and swine had all the same disease." The peasants, tending to these animals and trying to bury or burn the bodies, suffered from malignant pustule themselves.[48] Anthrax, normally present only at low levels, had exploded almost simultaneously across Europe. This happened again following a series of floods in the 1780s, when outbreaks of anthrax-like diseases killed large numbers of animals from Siberia (where 100,000 horses died) to France, and from Sweden to Italy. In 1793–1795, the disease was particularly severe in Bavaria and Lombardy after flooding.[49] "Anthrax years" depended on anthrax weather—and they both acted on "anthrax soil."

Physicians and natural historians became more and more interested in anthrax soils in the 1800s for several good reasons. Officials developing state sanitary and public health systems identified and attempted to control anthrax districts, and economic incentives loomed large as the scale of livestock production began to grow. Natural historians puzzled over anthrax as they studied the microbial world and tried to work out explanations for how diseases arose in animals and humans. Writing in 1874, Munich professor Otto von Bollinger noted that an excess of "moisture" and "decaying vegetable matter . . . appears to furnish the most favorable conditions of life for the poison [of anthrax]." Bollinger did not believe anthrax-like diseases were "malarial" or entirely due to the weather conditions, because the "poison" needed to be present in the soil for the disease to appear in a particular place. Rather, the weather created "a favorable condition for the preservation and development of the virulent material," and thus for outbreaks of disease.[50] The key to understanding anthrax-like disease outbreaks, Bollinger believed, lay in the soil itself. Only two years after Bollinger published his deliberations, an obscure German physician by the name of Robert Koch isolated and figured out the life cycle of the microorganism, designated *Bacillus anthracis,* which lived in the soil and was asserted by Koch to cause anthrax in otherwise healthy animals. This bacillus would become the material connection between the *champs maudits*—the cursed fields—and the outbreaks of disease that Fournier had classified as *charbon contagieux.*

For the next century, scientists continued to study the characteristics of anthrax-bearing soils, thus formalizing the traditional understanding of anthrax as a place-based disease. Anthrax districts were outdoor laboratories in which investigators could study, collect, and manipulate *Bacillus anthracis* and monitor anthrax outbreaks. In 1880, Louis Pasteur investigated the role of earthworms in transporting the infectious agent of anthrax from (mainly animal) gravesites to the surface of soil grazed on by livestock. Pasteur pointed out that the French region around the Aveyron River, home to calcareous clay soil, suffered the most from anthrax, whereas other areas with granite-based soils were spared. In his opinion, the locations of anthrax districts could be predicted by their predominance of neutral to mildly alkaline soils rich in nitrogen and calcium, such as those in the *département d'Aveyron.*[51]

Early twentieth-century surveys revealed these soil types in other known anthrax districts—the area between the Tigris and Euphrates rivers and parts of Europe. In North America, a broad belt of alkaline soils ran from Louisiana and Texas north through Saskatchewan, with another large area through the Pacific Northwest and California. This soil distribution coincided with the known anthrax districts of the United States, especially southeastern South Dakota, northeastern Kansas, southern Louisiana, and the delta in the great valley of California (the latter two areas also regularly imported infected animal products through major ports, and eroded soil from infected areas collected in the deltas). The distribution of anthrax-like disease outbreaks, then, depended in part on favorable soil conditions. As a map of anthrax outbreaks in the early 1950s showed, most counties in the United States that reported anthrax were in the anthrax districts—that is, the areas known to contain these alkaline, mineral-rich types of soils (figure 1.2). Early twenty-first-century ecological models of the United States show a definite conjunction between types of soils, the historical movement of cattle, and the presence of anthrax outbreaks, and this historical pattern has continued even as public health measures have made outbreaks less frequent.[52]

But what has been so important about soil composition? Did the *Bacillus anthracis* depend on a particular soil environment for its life cycle and interact ecologically with the components of the soil? Or did it simply sit in the soil, being tossed about mechanically along with other soil particles through cycles of wind, drought, and flood? This question about the role of the environment in disease transmission has remained surprisingly vibrant in the case of anthrax—surprising because twenty-first-century views of disease focus on what happens in the body, not in the soil. Indeed, most scientists throughout the twentieth century subscribed

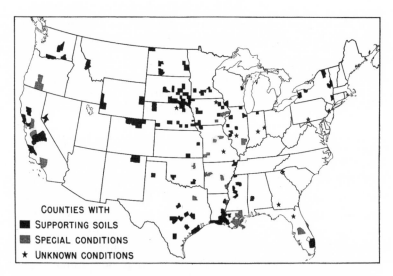

Figure 1.2. Map of United States counties with soils supporting Bacillus anthracis. Cases of anthrax in the United States usually appeared in these anthrax districts, which contained soils thought by some to support the growth and development of the bacilli. This theory was controversial because most scientists believed the bacilli grew only within human or animal bodies. Glenn Van Ness and C. D. Stein, "Soils Favorable for Anthrax," *Journal of the American Veterinary Medical Association* 128 (Jan. 1, 1956): 9.

to the "persistent spores" theory: *Bacillus anthracis* spores lie dormant in the soil; get into the ingested food, inhaled air, or open wound of a healthy animal; and only then can multiply within the protective environment of the body. This theory has also been challenged, however. Since at least the 1880s, some investigators have even wondered if specific soil attributes allowed the *Bacillus anthracis* to occasionally reproduce there, outside an animal body. This idea, that the soil enabled the microorganism to go through developmental stages necessary to foment an epidemic, can be traced to the German sanitary scientist Max von Pettenkofer (who had many followers in German and British public health). Pettenkofer, trained as a chemist and physician, was a proselytizer for fresh air, pure water, and sewage disposal systems, and he is generally credited with making sanitation a scientific endeavor. Famous for deliberately drinking a culture of cholera germs and not getting sick, Pettenkofer had complex views on disease causation. For our purposes, his most important idea was that some disease-causing microorganisms built up the virulence necessary to cause a disease outbreak while developing (or incubating) in the soil.[53]

Pettenkofer's incubation theory could help to explain why outbreaks of anthrax did not always correlate with the presence of bacilli: sometimes the disease did not appear when expected, and sometimes it seemed to appear spontaneously. This theory also explained why anthrax districts were characterized by certain climatic conditions and soils. An anthrax district could contain the bacilli for years, yet animals grazing on its pastures might not get sick. As we have seen, bad anthrax years often resulted when floods followed droughts in anthrax districts. According to Pettenkofer's theory, the floods did not simply stir up the soil and bring the bacilli to the surface, where susceptible animals could eat or inhale them. Rather, the climatic and soil conditions activated the microorganisms, stimulating them to develop or perhaps undergo portions of the life cycle. Therefore, the appearance of anthrax depended on the *Bacillus anthracis* interacting with its environment.[54]

From the 1860s through the early twentieth century, scientists interested in the environment's influence on disease outbreaks honed Pettenkofer's theory in order to find out which attributes of soils had the most influence on the persistence and virulence of anthrax (although they cited Louis Pasteur, not the more obscure Pettenkofer, as their prophet).[55] They measured soil pH, moisture levels, and numerous nutrients, attempting to correlate them with outbreaks of virulent anthrax. Proponents of the soil incubation theory included veterinarians, geologists, and biologists, and these ideas continued to be an important minority view, persisting through the early 1980s.[56] These scientists focused on what they called "incubator areas"—the right kind of soil type to host the *Bacillus anthracis,* seasoned by periodic flooding. United States Department of Agriculture veterinarian Glenn Van Ness stated

the position most clearly in a 1971 *Science* article: "Biological conditions have favored the multiplication and then sporulation of
the organism in [this] soil." As Van Ness first wrote in the 1950s,
he and his colleagues believed anthrax to be a "geophytic" infectious disease—not just inert but actively multiplying in the
soil.[57] Anthrax's ability to reproduce at low levels under certain
soil conditions would explain how outbreaks could occur in animals grazing on pastures that had been anthrax-free for decades.[58]
The incubator areas theory sought to create a new ecology for
anthrax, one that altered the traditionally maintained relationships between the parasite (the bacillus), the environment (the soil),
and the hosts (the affected human and animal populations). With
this type of thinking, Van Ness and his colleagues sought to "open a
new and practical approach to the biology of infectious diseases."[59]

This was more than an intellectual debate, since it affected
recommendations for the prevention of the disease. If the incubator areas theory was correct, then anthrax spores could probably
persist in inhospitable soils and weather conditions, but they
would not cause disease outbreaks. Only spores living in incubator areas could cause outbreaks. As Van Ness put it simply, "Livestock in pastures that do not have incubator areas do not develop
anthrax."[60] Using data from an outbreak in the United States in
1957, Van Ness and his colleagues noted a large number of disease-
free pastures that mapped exactly onto areas lacking the necessary
alkaline soil pH, high soil moisture, high nitrogen and calcium
content, and warm temperatures characteristic of incubator areas
for *B. anthracis.*[61] The ways to prevent outbreaks, then, included
altering the pH of the soil, draining it to dry it out, and carefully
selecting pastures for susceptible animals. Even pastures in known
anthrax districts could be made safe by altered fertilization patterns and careful drainage programs. This ecological framing of
the bacillus' life cycle contrasted greatly with the laboratory life of
B. anthracis.

Strains of *B. anthracis* entered the laboratory in the mid-1800s.
There, without soil of any kind, scientists and physicians observed
the bacilli completing quite a different type of life cycle—one that

required a milieu closely simulating an animal body. In the laboratory, the bacillus appeared very fragile unless in its spore form, and the spore form did not reproduce itself. Without its thick spore walls, the bacillus could reproduce, but in this form it proved to be easily killed off by competing bacterial species (common in soil), and it had specific nutritional requirements (those present in animal blood, essentially). Some scientists have even postulated that the bacilli require processes in the host animal's body to trigger their reproduction. Studying the bacillus' behavior in the laboratory has led to emphasis on the "mammalian environment," while studying the bacillus in the field highlights its activities in the soil.[62] Regardless of whether a given observer believes that the bacillus reproduced in bodies or in soil, *B. anthracis* appears to be a microorganism that must be understood ecologically in the field as well as intimately in the laboratory, since outbreaks of the disease are associated with specific soil and climatic conditions.[63]

INFECTIVITY, FEAR, AND CHANGE

Over the past 10,000 years, anthrax-like diseases have traveled with human populations and their domesticated animals, expanding from old areas to new. Anthrax was well known as an agricultural disease among laypeople and physicians alike, and they had definite ideas about anthrax even if they could not control it. The first and most important characteristic of anthrax was its infectivity from animals to people, and the second was its location in certain places. Once established in the soil of a new area (whether the Fertile Crescent or Barbados), the disease syndrome manifested itself in varied and complex ways, all of which motivated people to try to understand and explain it. Historical documents give us a feel for how humans perceived anthrax-like disease. From my research, one thing is clear: the single most profound reaction over time has been perpetual fear of potential outbreaks and downright terror when the signs of disease emerged. As in Griffith Hughes's account of outbreaks in the eighteenth-century Caribbean, acute anthrax appeared suddenly: it could kill livestock so quickly that it was (and still is) often initially confused with a lightning strike. The symptoms and the effects on the body were horrible; humans could get it from merely eating infected meat or caring for their animals, and anyone severely affected seemed to be cursed by probable death. No one knew for sure why an outbreak appeared and disappeared when it did (sometimes anthrax years failed to materialize even when anthrax weather occurred, and vice versa). All these attributes contributed to the fearsome reputation of outbreaks of malignant pustule or whole-body anthrax.

Anthrax is a disease because we have given it a name—a name that dredges up a deep agricultural legacy of experience.[64] The link between this constellation of signs and symptoms and the soil from which it came has changed over time, from the *champs maudits* to the incubator theory. For over a century, scientists have focused on the *Bacillus anthracis* as the material link between the disease and the environment. The story of interactions between people and anthrax, and especially how anthrax came to be a ubiquitous agricultural disease, cannot be complete without a realization that the life story of the bacillus has changed as well. The bacillus has spread to new areas of the world and adapted itself to new soils and places. Thus, geographic and mental maps coexist when we try to understand the biography of anthrax—the disease has a complex topography of infectivity and fear.

Historical accounts give us some idea of the power of this fear in guiding the response to an outbreak. In the 1720s, French physician Jean (Nicolas) Fournier recorded the reactions of local villagers to the development of an individual human case of charbon. The unfortunate victim was dragged off to die, alone except for a guard, in an isolated place far from the village. Even one case in a village inspired this draconian reaction as a way to protect the other villagers from a *charbon contagieux* outbreak, as Fournier attributed these actions.[65] This fear extended to threats to livestock economies. The Saint Domingue outbreak in the 1770s and nineteenth-century episodes of livestock anthrax in Germany, France, and Britain attracted national and international attention.

Governments sent physicians and natural historians to investigate and attempt to stop the epidemic. In the early 1870s, Prussian officials assigned Robert Koch the task of bringing a Milzbrand outbreak to an end, thus saving the local livestock economy as well as the lives of local people who were Koch's patients. This assignment gave Koch his entrée to experimental microbiology and initiated another chapter in the biography of anthrax: the career of *Bacillus anthracis* in the laboratory.

CHAPTER TWO

Availability

Understanding the Germ of Anthrax

❋ ❋ ❋

The first anthrax, as I have said, had previously been defined as a disease . . . Did the microbe exist before Pasteur? From the practical point of view . . . it did not . . . He shaped it.

Bruno Latour, 1988

It is possible, by exerting upon [the microbe] certain influences of which the experimenter is master and which he directs as he wishes, to rob it of its excess of energy and to make it, after diminishing its power to the degree necessary, no longer the agent of death, but that of preservation.

Henri Bouley, 1881

Repeated outbreaks of anthrax in livestock stimulated physicians and natural historians (people we would call scientists today) to try to understand and control the disease in France, Germany, Britain, and elsewhere in the eighteenth and nineteenth centuries. Researchers standardized the laboratory cultivation of what they suspected to be anthrax's causative organism, a rod-shaped bacterium named *Bacillus anthracis* by Ferdinand Cohn in 1875.[1] In the process, they elucidated its fascinating life cycle, studied how it caused disease, and used it as an exemplar for germ ideas and practices. They also began a long historical process of trying to alter the behavior, anatomy, and ultimately the genetics of *Bacillus anthracis*. These were critical steps in transforming anthrax from a soil-bound agricultural disease to an inhabitant of the human-built environment and a central character in bacteriology's intellectual history.

Understanding the cause of the disease anthrax has long been seen as a pioneering triumph of germ theories in the 1870s. As the story goes, once Robert Koch "discovered" the right microorganism, figured out how to grow it, and proved that it caused a disease recognized as anthrax, then the truth was known, and the world was grateful for it. At this point, I can envision Koch's ghost making impatient gestures and mouthing the German equivalent of "if only it were so easy." The process of understanding anthrax was messy, uncertain, and lengthy. Just figuring out which disease syndromes to include in the category of "anthrax" was problematic. Yet even with tentative agreement about what counted as the disease anthrax, there were several intellectual problems with determining its cause. The theory that bacteria cause diseases, so familiar today, had to be proposed, tested, and challenged. As historian K. Codell Carter has put it, "Causes are not discovered in the same way that one might discover shells on a beach."[2] While this process was playing itself out, natural historians found themselves in the uncomfortable position of having to advocate a particular hypothesis before they had any visible proof for it. Many had trouble believing that something so complex as a multisymptom disease could be caused by an organism that could not even be seen. There were practical problems, too: deciding how to describe, classify, and name the tiny organisms—and distinguishing the pathogenic organisms from the neutral or beneficial ones. Natural historians trying to prove that a microorganism caused anthrax sought to distill the essence of the disease in its victims in order to isolate some sort of causative "thing," and then to bring that thing into controlled conditions and figure out how to raise, breed, and control it. It took many people the better part of a century to define a causal connection between the disease anthrax and the *Bacillus anthracis* and to understand perhaps the most important aspect of anthrax's natural history: that the bacillus had to kill its host to complete its own life cycle.

Anthrax was a test case for the theory that a particular microbe causes a particular disease. Anthrax also famously played an important role in the early history of vaccine development, quite by accident. By 1880, physicians and natural historians were well into the process of creating the discipline of bacteriology, and the anthrax bacillus was one of the first organisms to be intensively studied.[3] In the pages that follow, I retell a familiar story—but with a twist. I argue that linking *Bacillus anthracis* to the disease anthrax, rather than being a sudden discovery, proceeded as a series of steps that centered on domesticating the bacillus to the laboratory.[4] By *domestication*, I mean bringing an organism into close relationship with human populations (often by way of animal populations).

First, natural historians had to somehow collect anthrax in the wild and bring it under their control, and the way they did this was by gathering samples of organs, blood, and other tissues from animals ill with or dead from anthrax. These samples were proxies for the disease itself—the material evidence that crossed the boundaries between the uncontrolled conditions of the field and the (somewhat more) controlled conditions of the laboratory.

The second step in this process was to isolate and cultivate the causative agent or agents of anthrax, if there was such a thing, in the laboratory. The best-known example is the work of Robert Koch, the German bacteriologist who developed principles of culturing bacteria and of linking them definitively with the diseases they caused (germ theories). In the late 1800s, germ theories reduced much of what physicians and their patients had thought about disease into one simple determinant: the presence or absence of a particular microorganism that caused the same disease in every animal it infected. Whereas the causes of diseases had long included climatic and environmental factors, the balance of fluids in the body, the inhalation of toxic vapors, and other varied elements, the theory of germs dictated that physicians and natural historians hunt for and find a causative microorganism. Besides Koch, legions of talented and well-trained physicians and natural historians worked hard to isolate organisms, understand their life cycles, and link them definitively to particular diseases. Koch used anthrax as one of his first test subjects in his work on germ theory in the late 1870s, and this work is an important part of our story.

Anthrax's domestication to the laboratory would not be complete without the final step: taking the isolated and cultivated causative organism (or organisms) out of the laboratory and seeing if it would infect animals under field conditions, then testing a preventive agent such as a vaccine. Louis Pasteur's public vaccination trials of anthrax in a sheep herd at Pouilly-le-Fort in 1881 are an excellent example of how a domesticated agent could be manipulated and, as the French microbiologist Henri Bouley put it, made to be the agent of preservation against the disease.

Bouley and other germ-theory advocates had their work cut out for them in getting the majority of people (or even the majority of natural historians) to believe that tiny organisms no one could see could put a strong man in his bed or kill a large animal such as a horse. But what this new causative theory of disease promised back in 1881 is well illustrated by the powerful statement by Bouley with which I began this chapter. If infectious diseases, the greatest killers in Bouley's time, could be understood, then they could be controlled. Control began with mastery of the microorganism domesticated to the laboratory to rob it of its killing power and to force it to become an ally of human disease-control efforts.

While those investigators who brought *B. anthracis* into the lab were able to control and manipulate it in countless ways, however, they were constrained by the characteristics of the bacillus itself.[5] For one, it needed specific conditions in which to grow, and other bacilli that looked similar to it could grow alongside it. In addition, after its host died, the anthrax bacilli seemed to disappear from the blood. Finally, it died out—did not complete its life cycle—if it did not kill its host (a fact that fascinated the early anthrax investigators). Bouley, Koch, Pasteur, and their many colleagues would probably have been shocked to hear that their work would provide a foundation for later uses of *B. anthracis* as a biological weapon, an agent of deliberate killing. In the late nineteenth century, the challenge lay with saving human and animal lives from diseases, although biological weapons were not unknown. This challenge promised hard work for all who studied anthrax and brought fame to very few. These physicians and natural historians developed standard procedures, ideas, and their own careers while creating practical solutions to the problems of farmers and livestock raisers.

WHAT CAUSED ANTHRAX?

In the 1800s, several theories of what caused anthracoid diseases, each supported by compelling evidence and ardent advocates, competed with one another.[6] Modern ideas of anthrax causation arose out of a vibrant theorizing tradition present since ancient times.[7] As we have seen, livestock raisers in the Western world had long understood anthrax as being associated with contaminated soils, "cursed fields," and certain climatic conditions—all essentially noncontagious causes. Soil and climate were the most important causation theories for outbreaks of anthrax among groups of animals, and various minor causes could also explain individual or unusual cases. Aristotle described a disease in a horse, perhaps anthrax or blackleg (an infection with *Clostridia* bacteria that causes severe muscle inflammation and death), which he believed was caused by the bites of shrews. For centuries, the swellings caused by blackleg were called "shrews" by livestock raisers, and the theory itself persisted into the eighteenth century.[8] Strange diets or gluttony could also cause anthrax-like diseases, according to Lucius Columella (first century) and Publius Vegetius (in his fourth-century guide to veterinary medicine, *Digesta Artis Mulomedicina*). As historian Jean Blancou has described, this idea was featured in the thirteenth-century writings of Franciscan monk Richard Rufus, who popularized Aristotelian theories through his lectures in Paris and Oxford.[9] Nor did causation theories centered on gluttony disappear in modernity. In 1869, the major German textbook of veterinary public health procedures (*Staatstierheilkunde*, or state animal medicine) attributed anthrax to overfeeding rather than contagion.[10]

Poisoning and the inhalation of gases from decomposing materials (also known as *miasmatic poisoning*) could, it was thought, also cause anthrax, often seen as happening in particular locations (cursed fields). Supporters of poisoning theories pointed to the speed with which anthrax attacked and killed its victims. In the 1780s, Jean-François Thomassin conjectured that the bites of poisonous animals caused the cases of *charbon malin* he observed in Burgundy, France.[11] Miasmatic poisoning—becoming ill from exposure to putrid vapors—also had its proponents. In 1673, a British animal doctor named Harward blamed the "infected air" of a fog for an outbreak of anthrax-like disease in cattle; in 1785, Joseph Enaux and François Chaussier wrote that the cause of anthrax in people was a "putrid entity originating from animals with malignant and anthracic fevers."[12] Louis Hurtrel d'Arboval, in his 1838 *Dictionnaire de Médicine,* suggested that spontaneous generation of anthrax infections could occur in the presence of sewer gas and fogs.[13] Thus, causation theories based on noncontagious sources predominated throughout much of the 1800s and continued alongside the slow development of germ theories in the second half of the century.

Contagion theories of disease causation had their proponents as well, since ideas of contagion had long existed in both popular and expert beliefs about certain diseases. In an 1875 review of the anthrax literature, Munich professor Otto von Bollinger cited Enaux, Chaussier, Nicolas Fournier, and others when he asserted that "the contagiousness of anthrax was proved" by the end of the 1700s.[14] Obviously, most people were not convinced in the late 1700s, in part because other theories were well entrenched and in part because poor microscopes meant that even most natural historians had never seen anthrax bacilli. Specific ideas about the mechanism of this contagion did not appear in the medical literature until well into the 1800s, and they did so in the context of debates about spontaneous generation, zymosis, and developing germ theories. Spontaneous generation referred to the idea that infectious particles were self-created. They could have their origins in predisposing conditions, but they were *sui generis*—arising in human and animal bodies without an obvious mechanism or portal of entry, and then causing disease. Zymosis referred to a process analogous to fermentation. Small particles got into the bodies of their victims and biologically transformed them. The infected body, as part of its putrefaction process, would then excrete poisons that infected other bodies. Several researchers, notably Charles Cagniard-Latour, Theodor Schwann, and Louis Pasteur, developed the theory of what Pasteur called "the theory of organized ferments," which inferred the life processes of minute organisms.[15] From the many types of zymotic theories, various natural historians developed versions of germ theories, all based on the idea that an infectious disease could not occur without the presence of a particular microorganism. The microorganism was the sole *cause* of the disease, regardless of other circumstances.[16]

Thus, through the 1700s and much of the 1800s, multiple causation theories for anthrax existed side by side. The circumstances of each individual case or outbreak suggested the cause most likely to be operating in that case.[17] Consequently, each causation theory suggested different experiments to test it, and these experiments slowly moved investigations of anthracoid diseases in a particular direction: to gather in the *essence* of the disease, reduce it to its simplest form, name it, and then begin to manipulate it. The way to understand the role of microorganisms in the disease was to bring sick animals and their microorganisms under the control of the researcher, and even into the laboratory. With magnifying lenses and the microscope, a whole new world of tiny organisms opened up to natural historians. They studied these tiny forms of life and their interactions with the much larger bodies of animals and people, in which they often lived. So many questions remained to be answered: What did these tiny organisms eat, and how did they live? What were their chemical properties, and how did they reproduce? Could any link exist between the presence of the organisms and the state of health or disease in the much larger bodies that harbored them?

In the 1800s, many people worked to understand anthrax. They were compelled by the deaths of people associated with af-

fected livestock and serious economic losses to undertake a systematic and professionalized survey of the several diseases that composed "anthrax" in the mid-1800s. Nowhere was this more pressing
than in France, where the cutaneous form of the disease (malignant pustule) had begun to appear in hide tanners and ragpickers,
as well as in farmers, butchers, and others involved in animal
agriculture. Beginning in the seventeenth century, French institutions in support of professionalized science had developed rapidly:
the Académie des Sciences (established in 1666); a thriving and
innovative system of academic training in hospitals; and journals
devoted to disseminating the results of scientific investigations.[18]
This approach was part of the French post-Enlightenment effort
to establish state-sanctioned studies of "scientific" animal husbandry, agriculture, and surveillance of disease problems.[19] From
the late 1700s to the mid-1800s, French investigators particularly
sought to build on their epidemiological knowledge about anthrax by exploring what might be causing it. They took the crucial
first step of taking samples from diseased animals and using them
as agents that could transmit the disease. The national French
veterinary school at Alfort was the center of much of the early
activity. In 1823, an Alfort professor, Felix Barthélemy, demonstrated that anthrax could be transmitted from sick to healthy
horses by way of contaminated feed and did blood inoculation
experiments.[20] Repeated outbreaks of anthrax in France and other
areas of Europe prompted both individual and institutional efforts
to understand the disease.[21]

The second half of the nineteenth century was a frenzy of researchers trying to identify "new" bacteria that could be causative
of disease. To do so, they first had to figure out ways to bring
the disease under scrutiny, and a common way to do this was to
study parts of bodies affected by disease. This tradition came from
the discipline of pathology, itself developing in the mid-1800s.
Physicians and proto-pathologists searched for answers in the
bodily material and architecture of disease victims, often postmortem. Anthrax particularly interested Pierre-François-Olive
Rayer, a prominent Parisian physician, who acquired some of

his anthrax-infected samples (animal blood and tissues) from M. Collignon, the inspector at the Monmartre abattoir. Rayer tried injecting healthy sheep with the infected blood. He was able to transmit this disease easily this way, and he carefully noted his postmortem findings after the sheep died. Rayer and his pupil Casimir-Joseph Davaine also traveled to the Beauce region of France near Chartres, a major anthrax district, to obtain fresh infected blood samples and inject healthy sheep with them.[22] Rayer concluded that these experiments removed any doubt that the "very energetic" infectious properties resided in the blood of infected animals; thus, the blood was a material conduit for the disease. Rayer considered his findings important enough to write up a description of the experiments for the Société de Biologie (Paris) in 1850. (The later controversy over who had "discovered" the anthrax bacillus would begin with this document.) Rayer offhandedly described small elongated bodies, about twice the length of the red blood cells, which did not move on their own.[23] These small bodies, often appearing in blood samples from Rayer's infected animals, were probably *B. anthracis,* although neither Rayer nor anyone else seemed to recognize them as significant at the time. Certainly, there was no sense that these organisms might be the cause of the disease or that they had any other particular significance; Rayer described them as he did all his other postmortem findings. Nonetheless, twenty-five years later, Rayer's associate Casimir Davaine claimed that he had been the one to notice the small bodies in the blood, and that he had sent the information to Rayer to publish—thus establishing Davaine as the "discoverer" of the bacillus.[24]

The French researchers were not the only contenders in the contest to identify microorganisms associated with anthrax—their German rivals were also keenly studying this disease. F. A. A. Pollender, a physician in Wipperfürth, claimed that he first took note of unusual small rod-shaped organisms (*stabförmiger Körperchen*) in the spleens of cattle that had died of anthrax in 1849 (a year prior to Davaine and Rayer). Curious about the fine rod-like bodies, Pollender studied them intensively in his rudimentary laboratory. He measured them, attempted to kill them with acids and bases, and found that he could stain the small bodies with iodine, making them easier to see. Pollender could not tell whether they were the common putrefactive bacteria or were definitively associated with anthrax, but he knew that these small bodies present in such massive numbers were not simply normal products of the cows' bodies. Pollender asked a range of questions: Were these small bodies the source of infection or merely the carriers of it, or did they have nothing to do with the infection in the animal's body? Pollender did not publish these results until 1855. Nonetheless, his work has been the basis for the assertion by German scientists and historians ever since that the anthrax bacillus was first discovered by a German.[25]

Precedence hardly matters, because both French and German educational traditions contributed to the rapid increase in observations and interpretations about the mysterious bacillus between 1850 and the 1870s. Next came experimental evidence that the disease could be transmitted by way of infected blood between humans and animals. Friedrich A. Brauell had been trained in Germany and Denmark but worked for much of his career as a veterinary professor in Dorpat, in what is now Estonia.[26] Brauell got a unique chance to study anthrax in human beings when an assistant at the veterinary school (who had worked with anthrax-infected animal carcasses) developed the skin carbuncles and died of the disease. Brauell found that injecting the man's blood into healthy sheep caused fatal anthrax in the animals. He noted the presence of the small bodies in the blood and stated that they had begun to move in the blood about three days after death (although these were probably not anthrax bacilli but the so-called bacilli of putrefaction). Interested in the range of the disease, Brauell injected other species, finding that among the domesticated animals, only dogs and chickens were relatively resistant to the infection.[27]

He also sought to isolate which of the small bodies was most closely related to the disease. Since filters with small enough pores were unavailable, Brauell tried an important innovation: he used the bodies of laboratory animals as filtration equipment. Brauell took blood from a pregnant sheep that had died of anthrax and took separate samples from the fetus. He found that the maternal blood, when inoculated into healthy animals, sickened them with the same disease that had killed the pregnant sheep, but fetal blood did not. Brauell reasoned that the placenta had acted as a filter, preventing the causative agent from reaching the fetal bloodstream. Robert Koch, who admired the ingenuity of these experiments, reproduced them with pregnant guinea pigs and mice and got the same results.[28]

At this point, the work of Brauell and his curious contemporaries did not always lead to a better understanding of the relationship between the bacillus and the disease. Brauell often seemed to confuse the bacteria of putrefaction with the similar-looking bacilli that we now call *B. anthracis;* he also at one point asserted that he had found the bacilli that caused anthrax in the blood of healthy sheep. Later, French historians eager to discount the work of German scientists claimed that Brauell's work had been a step backward.[29] On the contrary, Brauell's story illustrates the difficulty of the task for these early anthrax investigators. They lacked not only reliable conditions and practices for obtaining and studying the bacilli, but also an overarching theory of how the bacilli interacted with the bodies of the animals infected.

Just as Brauell's work was coming to a close, researchers in France were about to see a confluence of ideas that linked chemistry and biology, agriculture and human disease, in the quest to understand what microorganisms were up to inside larger animals' bodies. They shared a belief that a given microorganism was recognizable morphologically and was stable in its form, and that it had to look the same and behave consistently everywhere. Louis Pasteur was only the most famous of many workers who sought to provide evidence for these beliefs.[30] In the process, they worked hard to complete the second step of anthrax's domestication: cultivating the potential causative organisms in the laboratory. Among the French anthrax investigators, Alfort veterinary professor Henri Delafond most directly attempted to bring the mysterious bacilli into the laboratory. Writing in 1860, Delafond clearly saw the bacilli as a living, vegetative microorganism: he attempted to cultivate the small rod-shaped bodies and study their life cycle, speculating on the possibility that they arose from "seeds," or spores. As historian Jean Théodoridès has described, Delafond "envisaged the importance of *culture in vitro*" of the bacilli.[31] He encountered serious technical difficulties but came up with a couple of important conclusions: the bacilli appeared in the infected animal's blood only a few hours before death, they disappeared after death as the putrefactive bacilli appeared, and the bacilli seemed to multiply themselves in flasks of blood. Delafond remained mystified about whether the bacilli were a cause or a result of anthrax, but he believed them to be intimately linked to the disease.[32]

In 1861, Louis Pasteur published his groundbreaking work on the processes of fermentation, articulating a theory that explained a great deal about the lives of microorganisms. He laid out his theory of how yeast could live in the presence or absence of oxygen, calling the agents responsible for the chemical processes of fermentation "vibrions."[33] Casimir Davaine, who had not forgotten the small bodies he had first seen a decade before in the blood of anthrax victims, read Pasteur's article and felt that he had found a theory he could apply to the mysterious rod-shaped organisms. The bodies in the blood of anthrax victims presented a "great analogy of form with these vibrions," wrote Davaine. What if the organisms present in cases of anthrax were just as active, conducting processes similar to fermentation that affected the much larger body of the animal and caused the disease?[34] Davaine focused on anthrax and what he called the rod-shaped "bacteridia" for the next several years, work that linked the microorganisms more closely to the disease. Davaine believed that a given microorganism had a unique shape and appearance and was stable in its form.[35] In this view, anthrax bacilli (or bacteridia) had to look the same and behave consistently, whether they lived in a human or an animal, in Constantinople or in France.[36]

This point can hardly be overemphasized, because if the anthrax bacteridia were the same everywhere, then natural historians around the world could study them and know that they were contributing to the same body of knowledge and understanding some aspects of the disease in similar ways. In 1863, Davaine published his definitive description of the bacteridia (*infusoires filiformesque . . . des bactéridies*) that were, according to his studies, necessarily present in all cases of anthrax (*leur présence en sera la confirmation*).[37] He described the bacteridia as "free filaments, straight, inflexible and cylindrical" that manifested "absolutely no spontaneous movement."[38] This exact description would guide Americans interested in the bacillus five years later during an outbreak in factory workers in Walpole, Massachusetts. Physicians Silas Stone and Richard Manning Hodges and naturalist-physician Jeffries Wyman peered into a microscope to view bacteridia found in the blood of victims. Speaking before a meeting of the Boston Society for Medical Improvement in early 1869, Hodges began by describing the 1863 paper that Davaine had presented to the French Academy of Sciences and noted that Davaine's "views are generally accepted in France, especially in those parts where malignant pustule prevails."[39] Hodges sought to demonstrate that the organisms from Stone's patients were the same as Davaine's bacilli because they had the same shape, size, and movement behavior, and he relied on Davaine's careful description. While visual representations of the bacilli (drawings, lithographs, and eventually photographs) would become the standard by which bacterial identity was judged, Davaine's early descriptions of the bacteridia's morphology and lack of spontaneous movement remained crucial.

This research constituted the essential first steps in *Bacillus anthracis'* domestication to the laboratory, and it was not easy. The investigators worked in difficult circumstances. Davaine, for example, kept his experimental animals at a friend's farm and did his laboratory work in Rayer's rooms at Hôpital de la Charité in his spare time after completing his daily rounds. (He made his living as a practicing physician.)[40] Delafond had no facilities for sterile culture, yet he came close to cultivating anthrax bacilli in flasks of blood. For these and the other early anthrax researchers, the impulse was the same: gather and collect samples (blood, tissues) from animals ill with or dead of anthrax; then manipulate those samples in ways that would unravel the mysteries of what transmitted the disease. Focusing on the bacteridia, then, opened a new series of profound questions about the nature of these strange bodies. Were they living, and if so, how did they eat and reproduce? How could they be distinguished from other tiny organisms? In what ways did their life processes affect the health of the animals in which they lived?

DOMESTICATING ANTHRAX TO THE LABORATORY

Laboratory life for disease-causing microorganisms began with the idea of proving that these small bodies were in fact living things, akin to plants and animals. Agricultural analogies helped here: if something lived, it could probably be artificially cultivated. Jacob Henle, professor at Göttingen, Germany, in the mid-1800s, outlined similar thoughts about microorganisms in 1840. First, he separated the cause of the disease from the disease itself, arguing that the cause—the inducer of disease in the animal's body—was a type of parasite. Such small bodies could produce such a great effect only if they reproduced themselves, which Henle viewed as further evidence that they were alive. Finally, he wrote, "after it has been shown that the contagion is alive, there still remains the question of how the contagion works to bring about its damage. If it could be possible to prove that a contagion can be cultured outside the body . . . then such a contagion could only be a plant or animal." Culturing the contagion meant that it could be manipulated and experimented on—a possibility that took hold only slowly in the minds of most people studying anthrax.[41] The anthrax investigators first had to focus on the technicalities of cultivating the rod-shaped bodies outside their animal hosts (which proved very difficult to do).

In the early 1870s, Edwin Klebs, a student of the famous German pathologist, physician, and social activist Rudolf Virchow, followed Henle's principles and attempted to raise bacteria in his laboratory at the University of Prague. Klebs outlined three important principles necessary to defining a particular microorganism as the cause of a particular disease. First, the diseased organ or organs in the affected animal had to be carefully studied; then different types of organisms from the organ had to be grown and isolated from other organisms (pure culture); and finally, the same disease had to be produced in healthy animals by inoculating them with the pure cultures. Kleb's frustration lay with his inability to separate the different groups of organisms from one another. He could not fulfill his second condition of raising pure cultures.[42]

Nonetheless, Klebs, Delafond, and Davaine had all taken the important step of attempting to cultivate the small organisms present in infected blood and tissues. They chose anthrax for several reasons. First, it was a pressing disease problem at the time. Second, the rod-shaped organism and the disease had certain properties that made them easier to study. The bacteridia were quite large and easily seen under the microscopes of the 1860s, and they could be stained to enhance their appearance still further. The organisms could be reliably found in the spleen and, in the later stages of the disease, in the blood (although they quickly disappeared after death). Anthrax could be reliably reproduced in healthy animals by simple inoculation of bodily fluids from infected animals.[43] But anthrax had its inexplicable side as well. It was known to be associated with soils of particular places—anthrax districts. How could this organism live in both soils and animal bodies? What role did the soil part of its life play? In those cases where one animal did not give the disease directly to another, how was it transmitted?

These questions and more awaited Robert Koch, a promising student of Henle and his colleagues at Göttingen, who had taken a position as a district physician in Wollstein in the early 1870s. Koch, who trapped and bred experimental animals on his property and had a makeshift laboratory room in his home, began thinking about anthrax as a subject for his rapidly growing interest in the new field of bacteriology. Anthrax plagued the sheep (and some of the people) of Koch's district, and he did not lack for blood and tissue samples. He began studying anthrax in his spare time in 1873, after his wife gave him a good-quality microscope for his birthday.[44] Although biographer Thomas D. Brock calls Koch at this early period of his career "The Lone Scientist," Koch (and his biographer) would quickly acknowledge that much of the earlier German work on anthrax (and some of the French work, which he likely read secondhand) were available to Koch.[45] Edwin Klebs and Koch's teacher Jacob Henle, both of whom believed that living agents caused disease (Klebs passionately so), were also important influences on Koch. Klebs and Henle did not lack opponents in their beliefs, and Rudolf Virchow (probably the most influential German scientific thinker of his time) remained suspicious of this new theory of germs causing disease.[46] In this fertile intellectual atmosphere, Koch set about his own modest work by focusing on the major technical problems plaguing anthrax investigators who tried to cultivate the bacillus outside infected animals' bodies.

Koch proved to be a consummate technician, and given his modest means, he wisely pursued solutions to the technical problems of cultivating *B. anthracis*.[47] He created an innovative system of cultivation for *B. anthracis* that would become the basis of laboratory methods in the field of bacteriology, in particular developing ways to grow individual pure cultures of bacterial strains and to make microorganisms more visible by staining them. Koch insisted on absolute control of the microorganisms. Even when simply observing them, Koch *mastered* them. The famous hanging-drop technique invented by Koch for *B. anthracis* provides an excellent example. Not only did he need to encourage the bacteria to grow, but he wanted to watch what they did at all times in their life cycle. So he created an ingenious culture slide that ensured a fishbowl life for the microorganisms. Beginning with a microscope slide with a concave center, Koch topped it with a flat glass slip from which he had suspended a drop of fluid containing a fresh piece of spleen from an anthrax case. The fluid drop (aqueous humor from a cow's eye, which fulfilled the bacilli's nutritional requirements) hung suspended in the cave of the microscope slide.[48] The bacilli sat stationary in the microscope, warmed by a lamp—an environment that simulated the inside of an animal's

body. Koch could watch their every move. What he saw aston-
ished him—and eventually changed the course of medicine.

In their nourishing droplet, the bacilli answered many of
the questions that had frustrated Davaine and his fellow anthrax
investigators. How could blood that contained no rod-shaped
bacilli cause anthrax in healthy animals, as many had observed?
What could be learned in the laboratory about anthrax as a soil
disease? What accounted for the anthrax districts, the places that
seemed cursed by the disease, when other places not so far away
seemed free of it?[49] Koch reasoned that the bacilli's life cycle included "a proposed developmental stage," as yet undescribed, that
could explain all these observations. By watching the bacilli, he
thought he would be able to understand their life cycle. Peering
through the microscope, Koch saw to his surprise that the bacilli
seemed to do different things in different parts of the droplet preparation. Those closest to the center of the droplet had not changed
much. But as Koch's eye moved toward the edges of the droplet, he
noticed "bacilli that are three to eight times longer . . . display[ing]
a few moderate angles and bends." Even farther toward the edge
of droplet, the filaments got progressively longer and intertwined,
looking granular, as if they contained tiny dots. At the very edge
of the slide, where the bacilli were exposed to the most air, the
filaments "contain[ed] perfectly formed spores." In the micro-
world of this one droplet, Koch could simultaneously view "all the
transition stages" of the organism's life cycle, "from short rod-
shaped bacilli to long filaments containing spores." Finally, he
watched spores changing back into bacilli when he put them into
a fresh, warmed droplet of aqueous humor protected from the air.
After months of these observations, Koch had constructed a firm
vision of *Bacillus anthracis'* complete life cycle.[50]

The spores that formed along the edges of the droplet answered many of the longstanding troublesome questions. The bacilli did not disappear after the animal's death; they converted
into spores as soon as they hit oxygen with the loss of blood after
death. The spores were the form in which the bacilli survived for
so long in the soil. They could not be seen as easily as the rod-

shaped bacilli (particularly if one was not looking for them). Blood samples exposed to the air might have appeared devoid of bacilli but could nonetheless cause disease in healthy animals because they contained the all-but-invisible spores.

The spores provided the missing link that connected the soil-based disease with the bacilli in the living animal. Koch described what was already well known but now explainable: "Cadavers of animals that die of anthrax are sometimes buried or sometimes lie in the fields . . . Stalls and yards also contain blood and bacilli from diseased animals . . . mixed with the moist earth." The bacilli could survive in the moist environment, and "the germs of this parasite are deposited in the earth every summer."[51] Slowly exposed to air, these germs converted to spores and persisted through the winter and for some time afterwards. (Koch knew they remained virulent for at least four years.) The spores in turn entered a healthy animal's body through a wound and then caused the disease. Koch also suspected that animals got the disease after eating the spores as they grazed. He was unable to induce the disease by offering mice and rabbits spore-infected feed but acknowledged that other species might be infected in this way. (He also speculated about, but did not attempt, experiments on the ability of dust to convey the spores to the lungs.)[52] The completed experiments led him to the conclusion that "anthrax substances . . . can only cause anthrax if they contain *Bacillus anthracis* or its viable spores."[53] Koch had provided a framework of processes for cultivating and manipulating the bacilli and a model method for studying the life cycles of microorganisms associated with infectious diseases.[54]

Koch's manipulations not only revealed the secrets of the bacillus' life cycle but also initiated laboratory practices that would eventually alter its evolution. Like the other anthrax investigators, Koch got his original bacilli from farm animals. Thus, he began with types of bacilli that lived in his local environment, bacilli that had successfully completed their life cycles there. These bacilli were grounded in a particular place and in particular climatic conditions. Once they entered the laboratory, their place-based identity slowly eroded. The laboratory was meant to be a virtual space in which natural variation could be studied, manipulated, and even standardized. Koch scrupulously reported the details of his techniques because he expected others with similar equipment and in a similarly controlled environment to be able to replicate them, no matter their geographic location.

Materially, Koch's procedures probably contributed an accumulation of tiny changes to the bacilli themselves that differed from the changes they would have undergone if they had stayed in the pastures. In the laboratory, bacilli ate what Koch gave them and lived or died on his whim of providing the right temperatures and the right amount of air. But more importantly, Koch altered the material composition of the bacilli by repeatedly transferring them directly from one experimental animal to another. These were internal changes that he did not have the capacity to observe. He did look for visible changes, "to determine whether the bacilli change into another form after a specific number of generations," but saw none: "the results were always the same."[55] The bacilli were *monomorphic*—they did not shape-shift within animal bodies or turn into other types of bacilli over time. But Koch's actions of continuously inoculating the bacillus directly from one animal to another bypassed the spore phase of its life cycle and subjected it to intensive selection pressure over several generations. Evolutionary theory and present-day experimental evidence tell us that this process must have altered the genetic and developmental makeup of the bacilli that had been domesticated to Koch's laboratory. Koch felt it unlikely that "a mutation would occur" in the process of this inoculation series because he observed the bacilli reproducing by division and assumed that daughter cells were clones: "A few bacilli always generated significantly many more similar bacilli."[56] Even though he could not completely explain the phenomena he observed, Koch relished his glimpses into the private lives of the anthrax bacilli, generation after generation.

Koch's appreciation for the aesthetics of anthrax in all of its forms is particularly striking, and he took care with the representation of the bacilli in his publications (even exploring photography, which was still experimental at the time). To Koch, anthrax spores embedded in the filaments could "best be compared to graceful, artificially ordered strings of pearls." Cells containing the bacilli were "particularly beautiful" when plumped up with distilled water. Koch referred to the colonies and filaments he encouraged to grow as "beautiful." Individual filaments "united into the most delicate spiraling, twisting bundles . . . like a pile of glass threads."[57] In his next major publication on anthrax, Koch explored taking microphotographs of anthrax bacilli.[58] However, as he admitted, the long exposures required and the vibrations of the apparatus at first rendered the photographs too blurry to be published or "to be of use . . . as evidence of what one sees."[59] Instead, lithographic representations of Koch's photographs were reproduced in several places, including in the English translation of his work *Investigations into the Etiology of Traumatic Infective Diseases.*[60] The lithographs were an attempt to represent the activities of the bacilli that Koch had described so vividly. The bacillus' transformative and destructive abilities played out in these later representations, as the bacilli "grew," and "formed" spores with "infecting power," in John Burdon Sanderson's words (figure 2.1).[61]

For the skeptics (and there were many, including Virchow), Koch offered his methods in excruciating detail, described multiple repeated experiments, and cited the authority of eminent bacteria and plant expert Ferdinand Cohn, pathologist Julius Cohnheim, and their colleagues at the University of Breslau, to whom Koch had personally demonstrated his results over a period of several days. Cohn published Koch's written account of the anthrax experiments in the journal Cohn edited. Koch always felt that his published anthrax experiments were the first to establish a specific bacterium, *B. anthracis,* as the cause of a specific disease, anthrax—although Pasteur contested this.[62] Despite the controversies, Koch's techniques contributed tremendously to (slowly) changing ideas of disease causation, a new role for scientific experiment in medicine, and even new therapeutic directions.

Koch's anthrax work accounts for another important historical

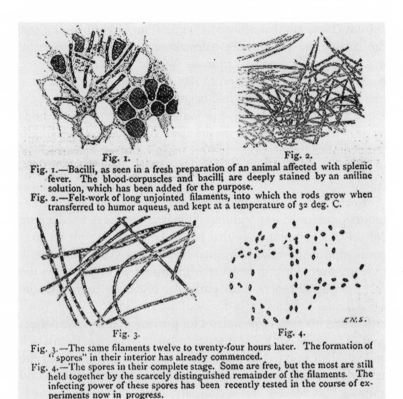

Fig. 1.   Fig. 2.

Fig. 1.—Bacilli, as seen in a fresh preparation of an animal affected with splenic fever. The blood-corpuscles and bacilli are deeply stained by an aniline solution, which has been added for the purpose.

Fig. 2.—Felt-work of long unjointed filaments, into which the rods grow when transferred to humor aqueus, and kept at a temperature of 32 deg. C.

Fig. 3.   Fig. 4.

Fig. 3.—The same filaments twelve to twenty-four hours later. The formation of "spores" in their interior has already commenced.

Fig. 4.—The spores in their complete stage. Some are free, but the most are still held together by the scarcely distinguished remainder of the filaments. The infecting power of these spores has been recently tested in the course of experiments now in progress.

Figure 2.1. Representation of the Bacillus anthracis life cycle, 1878. This vivid depiction of B. anthracis' life cycle, based on the lithographs of Robert Koch, make the bacilli into very active creatures. They are pictured here invading tissues, reproducing, and sporulating. John Burdon Sanderson, "Lectures on the Infective Processes of Disease," *British Medical Journal* 1 (Feb. 9, 1878): 181.

accomplishment: he had figured out how to bring *B. anthracis* into the laboratory, cultivate it, and force it into the various stages of its life cycle. He made this possible through his technical accomplishments: ascertaining that the bacillus could grow outside an animal's body in warmed aqueous humor; creating the hanging-drop, in which he could observe its actions; manufacturing glass culture chambers that allowed oxygenation without drying out the bacilli; and establishing painstaking procedures for serially inoculating experimental animals. Within a couple of years, Koch had identified how best to stain the bacilli and take microphotographs of them, and had begun using a plate-culture technique for cultivation along with flasks of liquid and animal bodies. Koch's elucidation of the bacillus' life cycle had profound public health implications, and his technical innovations made it possible for future researchers to manipulate the bacilli in the laboratory.

Koch's first paper said little, however, about how culture techniques produced a pure culture of the bacillus. He defined a pure culture as containing only that particular bacillus, with no contamination from other, unrelated bacilli. The goal of creating a pure culture mattered, because it identified an organism to the exclusion of all others. That particular organism could then be agreed on as the cause of a disease. Based on Henle's ideas, Koch articulated his famous postulates for proving that a particular organism caused a particular disease: isolate pure culture from a sick animal, then inject it into a healthy animal, see if it caused the same disease, and isolate it again in pure culture from the second animal. This process was technically difficult. Fluids from animals' bodies might contain many different organisms; how could researchers be sure that these contaminants weren't causing the effects they observed? This was a problem that had stymied Klebs and other researchers. Klebs tried a process that he called "fractional cultivation" (*fractionirte Cultur*) to convert the messy bodily fluids and samples into pristine laboratory cultures. He placed the samples or fluids into a flask containing a sterilized nutritive medium, let them grow, then took a tiny sample from this flask and inoculated it into another flask of sterile media. By repeating this process many times, Klebs obtained a fluid that contained only a tiny amount of the original infective organisms, and by inference, if those organisms had greatly outnumbered the contaminants, they would be the only ones to survive to the final dilution. He must have gotten inconclusive results upon inoculating his fractionally cultivated microorganisms into healthy animals, because he did not report his results in the literature; his technique had probably failed to produce the disease.[63]

There were other ways to use what one had on hand to raise pure cultures, however. In Britain, Joseph Lister created an ingenious apparatus that allowed him to isolate and count bacteria and even raise cultures from a single microorganism. He built what we might call a pipette: a syringe with a finely calibrated screw that could repeatedly dispense very tiny, known amounts of liquid. Using a micrometer, he could measure the diameter of a tiny drop on the microscope's glass slide, count the number of bacteria in the drop, and multiply by the known volume of the drop. By sequentially diluting the drop, he could isolate even a single bacterium, which he could then add to a sterile liquid to encourage its growth. If it grew, and retained its characteristics (bred true to type), he decided he had isolated a pure culture of a particular strain. The bacteria could then be studied and understood by how they behaved chemically as well as in the bodies of experimental animals. He named one bacterium that grew well in milk *Bacterium lactis,* for example.[64] Other techniques borrowed from beer-brewing practices, for which the researchers cultured yeast. Historian William Bulloch describes the ingenious development of glass culture tubes by E. C. Hansen, for whom the Carlsberg Brewery in Copenhagen served as laboratory. The tiny, slender tubes mechanically separated the yeast cells, which could then be removed individually by breaking the glass in specific places.[65] A multitude of microorganisms were subjected to these creative methods, and more, in the 1870s and 1880s.

The techniques of creating pure cultures in tubes, pipettes, and flasks (in vitro methods) were not the only ones that researchers attempted to use in producing pure cultures. Most researchers of the late 1870s, Koch and Pasteur included, believed (as Koch put it) "the animal body to be an excellent apparatus for pure cultivation" (this was known as an *in vivo* method). The two methods presented researchers with different challenges and rewards: in vitro allowed the researcher to watch and control more carefully (as Koch did with the aqueous humor preparation). But in vivo experiments could allow the researcher to test in vitro findings and to work with several different populations of bacteria at the same time. In an animal ill from an infectious disease, the observer could count on finding only one organism in large numbers, but experimental animals were also chambers for separating different populations of bacteria. [66] Koch explained, "If one injects blood containing only bacilli, these alone are transmitted, and they form a pure culture . . . if a field mouse is inoculated with both [micrococci and bacilli], the bacilli disappear and pure micrococci can be cultivated. It would certainly be possible to cultivate several varieties of bacteria together, separate them, and eventually combine them again. I regard the successive transmission of artificial infective diseases as the best and surest method of pure cultivation."[67] In France, Louis Pasteur believed that a healthy animal's body normally excluded bacteria and thus made a good vessel in which to culture homogeneous populations of *B. anthracis* (which is exactly what he did).[68] These techniques became standard practice, and anthrax bacilli became model organisms, eventually living in laboratories across Europe, Britain, and wherever the disciples of the leading institutions went. After Pasteur read and heard about Koch's work, he began intensively experimenting with *B. anthracis,* with his goal nothing less than harnessing and controlling its killing power.

MANIPULATING THE DOMESTICATED *BACILLUS*

When Robert Koch published his first anthrax paper in 1876, Louis Pasteur was fifty-four years old and already internationally famous for his work on fermentation, silkworm diseases, and spontaneous generation. Pasteur's laboratory life was very different from that of Koch at this time. Trained as a chemist, Pasteur had solved some pressing economic problems (*pébrine* of silkworms, potato blight, and spoilage of wine) and had received the gratitude of his government and the status of a minor hero in France. Pasteur had support: a proper laboratory, talented and devoted collaborators, and plenty of animals and assistants. He was also competitive, savvy, and an excellent public relations man. A particular aspect of Pasteur's work—the development of vaccines—was advanced in 1881 by a public demonstration of a new anthrax vaccine on a farm.

A vaccine is a tricky thing. It needs to be virulent enough to trigger the host animal's immune response, but it cannot be so strong that it causes debilitating disease or death. Making a vaccine requires harnessing the killing power of the microorganisms, titrating and controlling it, and being able to do so repeatedly. All doses of the vaccine need to have the same effect. In Pasteur's time, this was no easy task, given that one made a vaccine from living organisms. Living organisms possessed a range of behaviors and the capacity to change their forms and their environments. Nonetheless, Pasteur's background and experimental approach, and a considerable dose of luck, enabled him to manipulate and control the bacillus as no one else (save perhaps Koch) had been able to. Pasteur and his able associates altered the strains in his laboratory, sent the altered strains around the world by way of Pastorian vaccination institutes, and encouraged the commodification of the Pasteur vaccine.

Pasteur was not the only researcher seeking to manipulate *B. anthracis* in ways that would harness its killing power for the production of vaccines. Across the channel in Britain, John Burdon Sanderson, professor-superintendent of the Brown Institution, read and heard of Koch's work and brought *B. anthracis* into his laboratory. A major concern for Sanderson, and then for his successor, William S. Greenfield, was to ascertain what happened to the bacilli as they passed repeatedly through the bodies of rats. These investigators had two aims: first, to understand how they could modify the bacilli so that an animal would survive being inoculated with it; and second, to determine whether inoculation could protect the animal against a challenge with "unmodified virus."[69] Greenfield used Koch's aqueous humor hanging-drop culture method to cultivate the bacilli, and he made a crucial observation:

> When the Bacillus anthracis is artificially grown in successive
> generations in a nutrient fluid (aqueous humour) it maintains its
> morbific properties through a certain number of generations,
> but each successive generation becomes less virulent than its pre
> decessor, requiring both a longer time and a larger quantity to
> exert its morbific action; and after continuous diminution of
> virulence, at a certain stage in the successive cultivations, the
> Bacillus, though maintaining all its morphological characters
> and its power of growth, becomes completely innocuous even to
> the most susceptible class of animals.[70]

Greenfield carefully described how he had altered the bacilli.
He cultivated the bacilli in small closed tubes of aqueous humor
with which "every precaution" was taken to exclude other microorganisms. Keeping a sample of each culture, he then transferred
cultures to fresh tubes, growing in each successive tube a new generation of bacilli. A sample of each culture was also injected into
healthy experimental animals, and the first seven generations
killed the animals reliably. But beginning with the eighth generation, some animals began to survive the infection, and by the thirteenth generation, no animals died from inoculation. What was
happening to the successive generations of bacilli? Perhaps the
environment in which they were growing just didn't suit them,
what Greenfield called a "failure of transplantation." Perhaps they
could somehow reacquire the ability to cause the disease. Greenfield did not fail to note that whatever changes the bacilli had
undergone could be the basis of a protective vaccine, but he also
worried that opponents of germ theories would argue that if the
same bacilli did not always cause the disease, they could not *be* the
cause of the disease.[71]

Along with becoming superintendent of the Brown Institution, Greenfield conducted extensive experiments on anthrax in
1880 and 1881 to help answer these questions. During an outbreak
of inhalational anthrax in a Yorkshire wool factory, he obtained

fresh strains of bacilli taken from the bodily fluids of human victims. Greenfield used his findings as the basis for his 1880–1881
Brown Lectures, in which he declared the anthrax bacilli to be the
definitive cause of inhalational anthrax, or "woolsorters' disease."[72]
Greenwood's work got much less press than did the work of his
French counterparts, in part due to the obstructions of the anti-
vivisection laws and lack of funding (indeed, Greenfield left the
Brown Institution for a medical position in Edinburgh a year
later).[73] But clearly, in Britain as well as in France and Germany,
the *Bacillus anthracis* had become a major participant in the development of germ theories and germ practices.

This was the era of "comparative medicine," when veterinarians, physicians, and natural historians often worked cooperatively
on animal and human diseases. Henri Bouley, whose words began
this chapter, provides a good example. Bouley was a veterinarian
and professor at the Alfort veterinary school and a member of the
Académie des Sciences and the Académie de Medicine in Paris—
thus a member of both the scientific and the medical elite of the
nation. In the summer of 1880, Bouley paid a visit to a colleague,
Henri Toussaint, a professor of anatomy, physiology, and zoology
at the veterinary college in Toulouse. Son of a carpenter, Toussaint
had earned both veterinary and medical degrees and was a pupil
of the famous French pathologist Auguste Chaveau, another veterinarian who had done quite a bit of research in bacteriology.
Toussaint excitedly told Bouley about his considerable progress in
creating a vaccine for animals against anthrax. He made Bouley
promise to keep his accomplishments a secret until he could test
the vaccine preparation further, but Bouley, bursting with the
news, could not contain himself. Within the week, the Parisian
scientific community (including Pasteur) knew that Toussaint
had a promising vaccine. The question was how exactly he had
made it.[74]

Closely following Casimir Davaine's work from the 1850s and
1860s, Toussaint had set out to culture the bacillus and then to
manipulate it to create a type of vaccine. He first tried filtering out
the bacilli, then injecting the remaining liquid (which might con-

tain a specific poison that could vaccinate healthy animals) into his experimental animals. But this failed to protect them against anthrax, and he turned instead to heating the bacilli and exposing them to chemicals. By May 1880, Toussaint had decided that the best way to attenuate, or weaken, the bacilli was to heat them to 55 degrees Celsius for ten minutes in the presence of a low concentration of phenolic acid. He did not want to kill the bacilli; he just wanted to weaken their killing power while preserving their power to stimulate a healthy animal's immune system. After injecting these treated bacilli into numerous healthy animals, then exposing them to the disease, Toussaint was satisfied that he had found an effective vaccine. At the time of Bouley's visit, Toussaint did what any savvy and ambitious natural historian would have done, which was to send a letter with Bouley to the Académie des Sciences detailing his methods so that he could receive credit for it. Feeling that his results were preliminary, he sent the letter as a *pli cachete*—a sealed cover, dated but not to be opened until he had given permission to do so. When Bouley made the announcement in Paris, Toussaint had not yet given permission for the letter to be opened. He did so about three weeks later, when the letter was read before the academy. Pasteur and his associates learned that Toussaint had used heat and chemicals as attenuating agents. To make a field demonstration, Toussaint double-vaccinated about twenty sheep in a successful trial conducted at a farm belonging to the Alfort veterinary school. Toussaint believed that the attenuated bacilli left some kind of substance behind in the infected blood that protected the animal from a later anthrax infection.[75]

Thus, Toussaint published the first studies of acquired immunity against anthrax in animals, preempting Louis Pasteur. Nonetheless, Pasteur is widely remembered as the creator of the first successful anthrax vaccine. Why is Toussaint unknown today? Suffering from a severe neurological illness that began in 1881, Toussaint died at the age of forty-three in 1890. He was very junior to the eminent Pasteur when he conducted and published his important studies of anthrax between 1877 and 1880.[76] Toussaint also did not possess the inestimable advantage of assistants, stu-

dents, and institutional support that enabled Pasteur, the expert
marketer, to turn his cottage-industry laboratory into a widely
lauded global system of institutes. But Pasteur owed a great deal
to Toussaint.

Pasteur had created a successful vaccine against chicken cholera by exposing the cultures to the air for long periods, a method
that he announced publicly in October 1881. Without this exposure, the microorganism retained its virulence, making it useless
as a vaccine. Excited by the idea that this method could work for
other microbes, and hearing of Toussaint's work, Pasteur targeted
B. anthracis as the next microbe to alter in this way. His junior
collaborators, Emile Roux and Charles Chamberland (formidable
scientists in their own right), had been using a different method
of attenuating microorganisms. They found that exposing the organisms to chemicals in low concentrations could accomplish the
attenuation they needed to produce material for a vaccine.[77] Pasteur, Chamberland, and Roux ran *B. anthracis* through multiple
tests with heat, oxygen, and various chemicals. Following Toussaint's experiment with phenolic acid, Chamberland and Roux
tried adding potassium bichromate to the bacilli and made what
they considered the most reliable vaccine, one that would be
protective but not give a healthy animal the disease.[78] In the midst
of some early announcements about their anthrax vaccine, the
Pastorians received a challenge from another veterinarian who
wished them to prove the efficacy and safety of their vaccine publicly. Hippolyte Rossignol had a large farm in the commune of
Pouilly-le-Fort, a rich agricultural region, and he proposed that
Pasteur conduct a trial of the vaccine on sheep there. Farmers,
natural historians and physicians, government officials, politicians, and the news media were all invited.[79] Pouilly-le-Fort was
an international event. Not only were Pasteur and his vaccine on
trial, so were ideas about germs causing disease and how immunity worked overall.

Pasteur, Roux, and Chamberland scrambled to create a vaccine that *would not* fail. As Bruno Latour has described it, Pouilly-
le-Fort was a good opportunity for the Pastorians to take their

laboratory-based ideas, practices, and artifacts out to the field. They contracted with Rossignol on the terms of the trials, in effect trying to extend laboratory control over the field conditions, to "retain the balance of power," as Latour put it.[80] They needed to accept the challenge, but they also needed to mitigate the risk. Pasteur's vaccine, produced by exposing the bacilli to the air, was not performing as expected in the frenetic laboratory tests leading up to the field trials. It did not reliably protect animals from anthrax; some died. Chamberland and Roux saved the day by deciding that their much more reliable potassium bichromate–attenuated vaccine should be the one used at Pouilly-le-Fort.

For Pasteur, there were two potential problems with using this vaccine: this was not the type that he had led others to believe would be used, and Toussaint had already used a chemical (phenolic acid) to attenuate *B. anthracis.* Pasteur had wanted to be the first to use the technique that would prove successful; he did not want to share credit with Toussaint. When the animals were vaccinated at Pouilly-le-Fort, and while everyone waited anxiously for the results, only Pasteur, Chamberland, and Roux knew the method by which the vaccine had been produced. Their secret went to the grave with Pasteur and endured for a century.

In the early 1990s, historians Gerald L. Geison and Antonio Caddedu, working independently with Pasteur's original laboratory notebooks, uncovered what Geison called "the secret of Pouilly-le-Fort."[81] In February 1881, Pasteur had announced the method that had been used to make his new anthrax vaccine by hinting (although not stating) that it had been an oxygen-attenuated bacillus. He also took this opportunity to assert that the Pastorians' method was reliable, in contrast to Toussaint's "uncertain" method of attenuating the bacillus by heating it. In March 1881, Pasteur expanded on this theme by asserting that Toussaint had switched abruptly from chemical attenuation to heat attenuation only because he thought that was what Pasteur was doing. He cleverly associated Toussaint strictly with the heat method of attenuation (ignoring Toussaint's use of phenolic acid) and insisted that Toussaint's was an "artificial procedure" that would not create a stable strain of the bacillus for vaccination. In other words, Toussaint's science was flawed, and his vaccine was unreliable.[82] Toussaint, from his provincial and junior position, was unable to spar effectively with Pasteur. The illness that would kill him within a few years also began at this time. With the success of Pasteur's vaccine trials at Pouilly-le-Fort, Pasteur became internationally lauded as the creator of the first usable anthrax vaccine for animals. Historians have noted that this "undeniable element of deception" owed more to the importance of originality in scientific research than to any propensity for dirty tricks on Pasteur's part.[83] But it also meant that the Pastorians could control an internationally important business—the production and distribution of vaccines for the world's anthrax districts.

After the Pastorians' dramatic public vaccination demonstration at Pouilly-le-Fort in 1881, the development and use of anthrax prophylaxis became a matter of national pride. Indeed, as Maurice Cassier has shown, the Pastorians created a "*de facto* monopoly based on the industrial know-how needed for the production and quality control of the vaccine . . . and a system of contracts and exclusive operating licences granted to commercial enterprises and local agents."[84] Efforts conducted by Robert Koch's laboratory and the Hungarian government, among others, could not replicate the Pastorians' results in vaccination trials of animals; they alleged that Pasteur's published methodology was too vague. Restricted by French law from patenting a veterinary biological agent (such as a vaccine), the Pastorians kept critical details to themselves to maintain control over the production and mass diffusion of the anthrax vaccine throughout the 1880s and 1890s. Their product was widely used in France: around two million French animals were vaccinated between 1882 and 1887.[85] This is not to say that the Pastorian anthrax vaccine was not used in other nations. Pasteur and his colleague Chamberland oversaw the system by which laboratories to produce the Pasteur vaccine, which was "guaranteed" in quality, were to be set up in places like Budapest, Buenos Aires, and Madrid.[86] Elephants in the kingdom of Siam were to be vaccinated as well as sheep in Australia and New

Zealand, all through the intercession of local companies' formal contracts with the Pastorians.[87] Pasteur and his colleagues had successfully completed the domestication cycle for their strains of *B. anthracis:* they had brought wild bacilli into the laboratory, cultivated them, manipulated them to create new strains, and attenuated them to harness their killing powers. As Bouley had put it, they had "mastered," "robbed," and "diminished" the bacillus to change it from an "agent of death" to one of "preservation."[88] These accomplishments fulfilled two goals at once: one, to create tools that could be applied to problems of disease in human and animal populations; and two, to promote the germ theories and the methods and practices that supported this way of thinking.

Physicians and natural historians by 1881 were well into the process of creating what historian Michael Worboys has called "a research programme based on standard methods, an expanding group of practitioners, [and] cognitive exemplars"—the discipline of bacteriology.[89] *B. anthracis* was among the first microorganisms used in this discipline-building process. One important question, for example, was how scientists in different places could be sure they were studying the same organisms. Visual representations— drawings, photographs, and lithographs—of anthrax bacilli became the means by which scientists identified bacteria. Visual representations functioned as a "cognitive exemplar," or knowledge archetype, and they helped to defend germ ideas and practices against skeptics. This was especially important in the case of anthrax. For example, a Yorkshire physician named Edward Tibbets in 1881 challenged the link between the anthrax bacillus and the inhalational form of anthrax in people (woolsorters' disease) by focusing on William Greenfield's characterization of the morphology and behavior of the bacilli. "Might I be permitted most respectfully to ask Professor Greenfield to state, through your columns, the precise data upon which the identity . . . of anthrax has been conclusively established . . . what are the diagnostic characters of the organisms he so frequently designates 'typical bacilli,' and . . . what are the essential differences between the so-called bacillus anthracis, and the ordinary bacillus of putrefaction?" he queried the *British Medical Journal* in 1881.[90] Although Tibbets implied that many "gentlemen" would be glad to hear Greenfield's response, he was the only one who wrote in to question Greenfield, and Greenfield did not bother to reply.[91]

Instead, Leonard Cane, a Peterborough physician, wrote to gently admonish Tibbets that he needed "to consult the last volume issued by the Sydenham Society," Robert Koch's treatise on *Infective Traumatic Diseases*, where "he will find some very good plates [drawings] showing anthrax bacilli; also septicaemic bacilli."[92] As Greenfield's silence and Cane's rebuke indicated, physicians and natural historians increasingly found these visual representations of microorganisms' morphology to be authoritative evidence for germ ideas. The "very good plates showing anthrax bacilli" established the identity of *B. anthracis* and codified its relationship to the disease. The plates were meant to persuade (or perhaps silence) critics such as Tibbets.

Visual representations and textual descriptions of the bacillus attempted to transfer ideas about a crucial element of the *Bacillus anthracis'* identity—its "typical characters."[93] The bacillus' typical characteristics included both its appearance and its behavior. *B. anthracis* was shaped like a rod, was relatively large compared with other bacteria, and did not move. These attributes alone separated it from the more common bacteria of putrefaction, which did move in a drop of liquid. The anthrax bacilli were present in large numbers in animals in the acute stage of the disease, just a few hours before death, but they disappeared after death. The spores could be found in the area around the animal (in the grass or laboratory cage); when brought back into the laboratory and suspended in a drop of aqueous humor, they transformed themselves back into the bacilli. If the aqueous humor was sufficiently warm, the bacilli multiplied rapidly into long, distinctive chains. *B. anthracis* could also be distinguished from other microorganisms biologically, by its reactions to chemicals, growth media, and the manipulations of cultivation.[94]

Thus, in the late 1800s, the disease anthrax became in some ways synonymous with the presence of an identifiable organism,

Bacillus anthracis. This connection had different advantages for different observers. For natural historians, the specific attributes of the bacillus, and its life cycle (including the manifestations of the disease), demonstrated the bacillus' past interactions with its environment and how it could adapt to challenges. For physicians, *B. anthracis* was a *specific* cause of anthrax, and the problem of anthrax outbreaks needed to be addressed by controlling the bacillus. *B. anthracis,* even in tiny numbers, attacked a healthy animal or person and consistently caused similar pathological changes in all its victims. Indeed, to be considered a member of the *Bacillus anthracis* species (as opposed to *Bacillus cereus* or other closely related bacilli), a bacillus had to prove that it could cause anthrax and could kill experimental animals. In a very real sense, *B. anthracis* was defined by its ability to kill, indeed by the necessity of killing its host to complete its life cycle.

THE BACILLUS AND THE DISEASE

A century after the events I have just described, the French philosopher Bruno Latour penned the quotation that began this chapter. "The first anthrax," he wrote, "had previously been defined as a disease." This implied that the process of domesticating the bacillus in the laboratory was a kind of watershed in the history of the disease, and that the disease Latour called "first" anthrax was somehow different from the entity linked to the *Bacillus anthracis* ("second" anthrax). To some degree, the historical evidence supports this idea. People identified first anthrax according to its appearance in certain places, environments, and soils, and according to well-established patterns of behavior in affected animal and human populations (malignant pustules, sudden death, etc). It certainly was a disease rich with meaning.

But so was second anthrax, and here is where we need to proceed with caution. The identity of second anthrax, Latour implied, was reduced to being determined by the presence of the *Bacillus anthracis.* The historical evidence does not support this conclusion, as we will see in the next chapter, in which bacteriologists struggle to identify and understand outbreaks of inhalational anthrax. This form of the disease was confusing because often the sufferers did not have the skin pustules that were so characteristic of anthrax. The anthrax bacilli might be present but be mistaken for other bacilli that looked similar; or the scientists might have been looking for them at the wrong stage of the disease (when they were not evident yet in the blood). Or perhaps, thinking that the inhalational anthrax was a type of pneumonia caused by other bacteria, scientists might have ignored the rod-shaped *B. anthracis* if few were present. It was much easier to find the bacillus if a physician or bacteriologist saw the skin pustules and knew what he was looking for.

Even if the disease "anthrax" had become synonymous with the presence of *Bacillus anthracis,* the very understanding of the bacillus and its properties was the product of a historical process. "Did the microbe exist before Pasteur? From the practical point of view," Latour concluded, "it did not." *B. anthracis* itself had been constructed as the cause of the disease by natural historians and physicians such as Pasteur. Understanding the bacillus required the talents and hard work of many people. The important point here is that the process of domesticating the bacillus (accomplished by Davaine, Brauell, Koch, Pasteur, and their colleagues) emerged from what these people understood about first anthrax. The very process of constructing *Bacillus anthracis,* and thus the way scientists understood the bacillus, could not be divorced from the experiences of cursed fields, sudden deaths, drops of aqueous humor from a cow's eye, and the observed reactions of animals' bodies. So much of what people knew about first anthrax was incorporated in how they understood the bacillus that it is impossible to separate the two. Thus, the transition from first to second anthrax was a smooth continuum, a road that curved. It was not the replacement of first anthrax by something entirely new or a revelation of a truth about the disease. This is one important conclusion we can make from understanding that *B. anthracis* was domesticated, not discovered.

The process of domesticating *Bacillus anthracis* did establish some important ideas and procedures for bacteriologists. The technical difficulties of creating pure cultures, for example, stimulated various researchers to try glass tubes as separators and animal placentas as filters. However, domestication was a different process for every microorganism precisely because so much of what people knew was unique to that organism and the diseases or effects that it caused. As Worboys has remarked, "it would be wrong to imagine . . . an 'anthrax analogy'" that was the model for the discovery of all diseases' causative organisms.[95] Rather than the anthrax studies being some kind of recipe for studying other microorganisms, they can signify the complexity of the process of domestication.

One important result of *B. anthracis'* domestication process was that it made the bacillus a common denizen of bacteriological laboratories and institutions almost from their inception. Throughout the late nineteenth and early twentieth centuries, the training of young bacteriologists (and pathologists, veterinarians, and physicians) included culturing *Bacillus anthracis* and its relatives in the laboratory. This seems strange to us today, given *B. anthracis'* propensity to cause a fatal disease. But anthrax was still a common agricultural problem and thus was likely to be seen in the field or in medical practice. Students needed to know how to handle, identify, and manipulate the bacilli. Moreover, historical precedent counted for a great deal, and students worked on the organisms that had brought fame to their predecessors (especially in the French and German bacteriological institutes founded in honor of Pasteur's and Koch's accomplishments).

Thus, by the beginning of the twentieth century, much was known about the organism, and many people knew how to cultivate it. This made *Bacillus anthracis* available for consideration later as a microorganism that could be made into a biological weapon. As the American microbiologist and biological weapons advocate Theodor Rosebury put it in the 1940s, "If an agent is to be used . . . the agent itself must be at hand, and it must be possible and practicable to prepare enough of it to meet its particular requirements."[96] While it was not inevitable that *B. anthracis* would become a biological weapon, the bacillus was certainly available by virtue of its long history in the laboratory.

Another important conclusion is that both parties—the scientists and the bacillus—got something out of the domestication arrangement. Scientists enhanced their own careers while creating practical solutions for agricultural problems. The bacillus acquired a new habitat and new territory, and likely increased its genetic diversity through its continual movement in vitro and through the bodies of laboratory animals. Although investigators who tinkered with *B. anthracis* in the laboratory sought to create pure cultures, ironically they were unwittingly altering the bacilli and creating more diverse populations of microorganisms. The range of *B. anthracis'* biological diversity stood to increase steadily simply as a result of the common procedures used by anthrax investigators. For example, Davaine, Koch, Pasteur, Greenfield, Toussaint, and many other researchers relied heavily on the technique of serial cultivation, or *passaging*. Passaging, defined as the inoculation of the bacilli into a succession of culture media or animal bodies, forcibly altered the bacilli (although the researchers did not know how or to what extent). Those bacilli that exited the animal's body were subtly different from those that had first infected the animal.

Techniques such as passaging served the interests of natural history because they helped to deconstruct the inner workings of the bacilli and to measure responses to changes in the environment. In medicine, passaging altered the ability of the bacilli to infect bodies and cause disease or death. A physician or natural historian could manipulate *B. anthracis* into becoming less virulent and thus perhaps useful as a vaccine or preventive treatment. She or he could also attempt to increase a pathogen's virulence. Either way, passaging altered the bacilli in essential ways that late nineteenth-century scientists could not yet elucidate, creating something slightly different with each isolatable generation of *B. anthracis.*

Even as physicians and natural historians tinkered with the

bacillus, they were uncovering more information about the disease. Despite the earlier work of Jean Fournier, Philibert Chabert, and others who had tried to classify the types of anthrax definitively, all the different signs and symptoms of the disease still needed to be related somehow to the bacillus. Far from spending all of their time in the laboratory, anthrax investigators in the late 1800s found new sites at which to collect and study anthrax and the associated *Bacillus anthracis.* Creating knowledge about anthrax, and continuing the ongoing process of domestication, led the anthrax investigators to factories, fields, and stables—and a new global understanding of the disease's ecology.

CHAPTER THREE

Transmission

Anthrax Enters the Factory

❋ ❋ ❋

R. N.—, aged fifty-eight years, married, strong, stout and healthy-looking; had been a [wool]sorter over forty years. . . . He had great difficulty in walking home, so complete was his exhaustion . . . he said he felt "very poorly," "completely done," but had no pain . . . [the] pulse [became] very rapid, irregular, and uncountable; hands, knees, face and tongue cold . . . mind clear. He died three hours afterwards.

One of seven cases described by John Henry Bell in 1880

These poignant medical notes provided a stark portrait of one wool-mill worker's encounter with *Bacillus anthracis*. The likely origin of his disease was, strangely enough, an infection in goats thousands of miles away from his home. This case, and hundreds of others recorded over the past 150 years, reflects a major transformation in the identity and global ecology of anthrax. *Bacillus anthracis* greatly expanded its habitat, becoming established in manufacturing centers in the early 1800s because it was carried around the world by the trade in wool, hair, hides, and other animal products. This trade contributed to an ecological exchange of anthrax strains between the Far East, the Middle East, South America, Europe, and the United States. Areas in Turkey, Siberia, Argentina, Peru, and Chinese provinces sent wool, hair, and hides from native animals to the factories of Western Europe and North

America. Yarn, textiles, and leather traveled between nations with unprecedented speed, carrying the footprints of the bacillus: Parisians wearing overcoats were closely connected to nomadic Levantine herdsmen, Yorkshire wool sorters, Turkish goats, and Siberian ponies through *Bacillus anthracis* spores retained in the wool. This ecological exchange had important ramifications for the people who produced and consumed these products, and for the *Bacillus* that rode along with the goods.

B. anthracis was an important organism in the development of ideas and practices in early bacteriology. At about the same time, the increasing incidence of malignant pustule and woolsorters' disease in industrial workers made these bacteriological investigations more than a minor scientific curiosity. In the 1870s, several dozen workers in Yorkshire factories that imported *B. anthracis*–contaminated wool died of a strange and frightening disease. Most were from in and around the mill town of Bradford. Like the case of R. N. at the beginning of this chapter, some victims felt fine one day and died the next, others developed fevers and skin lesions (malignant pustules), and only a few survived (there was no effective treatment). The majority suffered severe pneumonia-like symptoms, exhaustion, fever, and collapse—and not the skin lesions that were suggestive of anthrax infection. Thus, during the outbreaks of the 1870s, no one knew that the mysterious disease was what we would call the inhalational form of anthrax. The disease was simply named for the human population it most affected: woolsorters' disease. This chapter takes us back to the 1800s so that we can see how "woolsorters' disease" became "anthrax." Painstaking bacteriological and epidemiological investigations, especially the one initiated by physician John Henry Bell at Bradford, transformed woolsorters' disease into inhalational anthrax by 1881. These investigations linked the two forms of anthrax and demonstrated how bacilli could be traced in contaminated products shipped around the world.

By the 1880s, woolsorters' disease was common enough in Bradford that the French called it *maladie de Bradford*. Perhaps this was meant to divert attention from the fact that French workers suffered from woolsorters' disease, too, in places such as the mill town of Montpellier, for example. German physicians knew of the disease, and no doubt it appeared in other industrialized places as well. Although inhalational anthrax was most closely associated with spinning mills, the disease no doubt appeared sporadically in regions around the world where people raised sheep, goats, cattle, and horses. People who worked with livestock mostly suffered from the malignant pustule form of the disease (the blackened skin swellings), which was less often fatal.

Anthrax was not, however, a common agricultural disease everywhere. In Yorkshire, for example, the disease was quite rare in native cattle and sheep and in those people who handled them. It took British anthrax investigators quite a while to connect woolsorters' disease with the malignant pustules of anthrax because the agricultural counterpart of the industrial disease was quite rare in Yorkshire. Even if the local livestock had suffered intermittently from the disease, there would have been little reason for physicians and natural historians to connect it with the strange pneumonias suffered by the mill workers. In other words, no obvious direct causal connection existed. Anthrax was thought to be agricultural and local to the anthrax districts of Europe and the Middle East.

For these reasons, and because much of the early scientific work on *B. anthracis* was being done on the continent, British and American physicians were caught by surprise by the outbreaks of woolsorters' disease in the late 1860s and 1870s. Woolsorters' disease had been reported only a few times in the past, and few physicians had seen cases of it; thus it took them some time to suspect anthrax. To do so, they used an interesting combination of bacteriological techniques (developed in France and Germany) and ordinary detective work (interviewing workers, for example). The process was difficult and included wrong turns and complications. At times, these investigators suspected that they were dealing with anthrax, but mostly they found the workers' symptoms and the patterns of transmission to be very confusing. Once they caught on, however, British and American researchers were at the forefront of developing methods for investigating and attempting to control industrial anthrax. Regardless, the global trade in bacilli-contaminated animal products (usually raw wool or fleeces) spread strains of bacilli well beyond their native locations. Faster ships, cheaper transport costs, and reduced tariffs on imported goods all contributed to the increase in global trade—and thus the increase in woolsorters' disease in mill workers.

These changes in global trade had a heavy impact on Bradford, which was the worsted yarn-spinning capital of the world in the mid-1800s. Worsted yarn differed from woolen yarn because its fibers ran evenly parallel in the yarn strand. It generally required that fleeces be sorted (thus the job of wool sorting) before being spun. Around 1800, inexpensive and abundant long-stapled fleeces from Asian sheep and goats began to be imported into Bradford district by mill owners such as Titus Salt. In 1837, Salt stood at the pinnacle of England's worsted industry. Head of a well-known family, he controlled an entire town of mill workers—Saltaire, located on the lonely moors also inhabited by the Brontë family of literary fame. Titus Salt took pleasure in continual innovations in his textile mills, including searching the world for new sources of cheaper raw materials to meet the ever-increasing demand for worsted yarns. By 1840, Salt and his technicians had integrated two new types of fleeces into their mills: goat mohair from the nomadic herds of Asia Minor (imported from Constantinople) and alpaca from South America. At Saltaire, the use of mohair and alpaca increased greatly between 1840 and the 1880s, spurred on by the development in 1850 of a mechanical wool comber especially suited to these long fibers.[1]

During this time, an increasing number of Saltaire woolsorters, the workers who opened and sorted through the bales of imported fleeces, became ill with a puzzling disease that left them coughing, febrile, gasping for air, and usually dying within a few days. Local physicians were divided on what could be causing the disease, although it seemed to be some kind of poison from the wool-sorting rooms that quickly overwhelmed the bodies of even the strongest people. The workers, frightened by the acute and often fatal nature of the disease, made their own conclusions and began refusing to work with some materials from particular parts of the world, notably the Turkish goat mohair. By recognizing the disease as associated with something in the raw materials, and then associating it with geographic locations, mill workers in Saltaire and other areas forecast a new identity for anthrax as a globally distributed, industrialized disease (although it took some time for this belief to catch on with physicians, mill owners, and public health officials).

By the 1880s, anthrax came to be domesticated in factories that used raw animal fleeces and other products, and in the process this agricultural disease gradually became an industrial one. Once anthrax spores were embedded in the bricks and mortar of a factory building and in the bodies of the workers, the spores were permanent residents. The spaces inhabited by anthrax expanded in concert with the development in several nations of what historian John F. Richards called "a true world economy" as well as urbanization and industrialization. The new anthrax towns joined the old rural anthrax districts.[2] Mills that used wool, hair, and hides to manufacture yarn, mattresses, and shaving brushes played major economic roles in these towns but also became centers of mysterious afflictions affecting the workers and eventually the local livestock. This process signaled anthrax's escape from being literally grounded in local geographic districts. Now the disease could be found wherever infected fibers could travel. With the increase in global trade, based in trading centers such as Constantinople, local strains of anthrax became portable. Twenty-first-century genetic profiles of anthrax strains support the idea that strains from around the world hitched a ride on animal products (such as horse hair, hides, and wool) in the 1800s. A Siberian or Armenian agricultural infection could become established in a German, English, or French factory town; and the processors of raw fleeces in Constantinople could suffer the same infections as Yorkshire woolsorters. For interested scientists and physicians, the factory itself eventually became a laboratory: a source of ideas, spores, patients, and test populations for vaccines. It also gave investigators an opportunity to study closely the inhalational, or pneumonic, form of anthrax, which would be the form of the disease most closely associated with the use of anthrax as a biological weapon in the twentieth century.

"MELANCHOLY HISTORIES": MECHANIZATION AND WOOLSORTERS' DISEASE
The Wool Trade and the Mechanization of Wool Spinning

In the early nineteenth century, the textile manufacturing industry grew tremendously in the United States, Britain, France, and the Netherlands. By no means was this a new industry, of course; around the world, making cloth was a labor-intensive part of most families' daily lives and even generated income for those families able to make more than they needed for themselves. In the mid-nineteenth century, the beautiful, rugged county of Yorkshire, England, was the center of the world's worsted and wool industry by dint of bountiful local labor, coal, sheep, and British industrial development. Citizens of Yorkshire towns migrated around Europe and across the oceans, where they used their skills and tools to re-create their wool-producing economy in places like Mexico, Argentina, South Africa, Australia, and New England. Yarn and textile production, formerly a very local, family-based activity, became industrialized in the late eighteenth century, spread to new geographic centers, and grew rapidly in the early 1800s—so rapidly that local wool supplies had to be supplemented with materials imported from around the world.[3] In the large mills, woolsorting was considered one of the best and healthiest jobs. Woolsorters were skilled workers: usually literate and well respected, they commanded higher wages and formed their own benevolent societies. They would prove a resourceful and articulate group willing to agitate for public health reform when occupational disease struck.

The larger-scale importation of fleeces in the mid-nineteenth century built on a longstanding tradition of bringing fleeces across long distances. In earlier times, production of wool yarn and cloth had not necessarily been located near large groups of wool-producing animals, and trade for wools flourished. The famous Flemish wool markets, prior to Spanish conquest, were the center of Europe's trade, while Mexican artisans worked vicuna and alpaca wools from South America.[4] Worsted yarn production in Yorkshire became industrialized between about 1790 and 1850, and a few large mill owners consolidated their wealth and power during this period. By the 1870s, Yorkshire began to share its dominant position in textile production with France because the demand for worsted products changed, but for much of the 1800s, towns such as Bradford, Leeds, and Huddersfield owed their social structure and institutions to the worsted factories.[5] From Yorkshire, the industry spread to other areas. Mexico began industrializing its textile industries with British machinery processes between the 1820s and 1840s, and a major center of machine wool milling (based on immigrating Yorkshire workers) developed in New England in the United States.[6] In the mid-1800s factories and mills in these areas had all but supplanted local home production of "common" products.[7]

The upsurge in cases of industrial anthrax in mid-nineteenth-century Yorkshire likely lay with the increased use of large amounts of imported fleeces in the mills. The numbers of bales of mohair imported to Britain from Asia Minor—the type of fleece most often associated with early anthrax outbreaks—increased almost tenfold between 1840, when Titus Salt began using it at Saltaire, and the 1880s.[8] In turn, Yorkshire's demand for mohair influenced its production in Asia Minor. At the beginning of the nineteenth century in Asia Minor, finely bred Angora goats produced a high-quality, long-stapled mohair fiber. As demand built in the 1840s through the 1860s, the Angora goats were increasingly crossbred with common goats to increase production. It has been claimed that the pure-blooded Angora goat had disappeared in its homeland by 1863, and along with it went the very high quality fleeces.[9] Thus, industrialized textile production and global trade had changed the animal population itself, as well as the raw materials sent to England's sorting and spinning mills. The cheaper, inferior Van mohair, from crossbred animals in the Lake Van district, was scorned by workers in Bradford, the destination of most mohair imported into England.[10]

As demand from the mills increased over time, livestock producers increased their output and sought to expand their slim profit margins by cutting fleeces from animals that had died. These fleeces were commonly referred to by the workers as "fallen fleeces," and the fibers were often matted with blood or still had bits of skin attached—prime locations for anthrax spores to linger. Another mid-nineteenth-century development ensured that raw goat hair could be most profitably used: the horizontal circular wool comb, also known as the Lister comb, mechanized the combing of mohair, alpaca, and other long wools. This machine became an industry standard in the Bradford area around 1850, and it contributed to the even greater demand for (potentially infected) imported mohair in the ensuing decades.[11] Because of increased demand, Bradford manufacturers found ways to use dirtier and more inferior—but cheaper—fleeces. As the workers surmised, the mills' increasing reliance on these new types of raw materials coincided with a dramatic increase in cases of woolsorters' disease in the 1870s.

Incidence of Woolsorters' Disease before the 1870s

Woolsorters' disease was probably infrequent before the mid-1800s, but it did make occasional appearances among factory workers in the industrializing areas of the world. In the 1750s, French physician Nicolas Fournier discussed the vulnerability of French wool and hide workers, analyzing outbreaks in the Montpellier tapestry mills.[12] The skin form of anthrax, malignant pustule, appeared more often than did woolsorters' disease in the early years. Fournier's colleague Monfils also described an outbreak of malignant pustule in 1776; Rudolf Virchow, in his influential "Milzbrand und Karbunkelkrankheit" ("Spleen fire [anthrax] and carbuncle illness"; 1855) cited cases reported by Armand Trousseau in the 1840s.[13] From these and succeeding accounts, it is clear that the workers themselves knew about the various forms of the disease and associated it with particular raw materials. French scientists, surgeons, physi-

cians, and veterinarians such as Pierre-François-Olive Rayer, Philibert Chabert (the principal of the Alfort Veterinary School), and
Alexis Boyer asserted the infectiousness of the disease early on;
their work appeared, translated, in journals throughout Europe
and in the United States. Textbooks as well as journals transmitted
French ideas to physicians even in provincial places. When an
American physician, C. W. Pennock, published one of the few
early accounts of North Americans with anthrax in 1836, he cited
Boyer's *Traite des Maladies Chirurgicales* as his major source of
information about the disease.[14]

Although French ideas predominated in much of the world,
German investigators had also published a great deal of work on
agricultural anthrax and were aware of its industrial implications.
By the mid-nineteenth century, German ideas were accessible to
other investigators mainly through translations in a few widely
available textbooks. The leading German textbook of surgery,
J. M. Chelius's *Handbuch der Chirurgie,* taught German students
that anthrax "may take place during the preparation of wool,
hides, and so on . . . the malignant pustule is therefore most
commonly observed in butchers, tanners, woolbeaters, [and]
shepherds."[15] Munich professor Otto Bollinger's informative and
comprehensive article "Anthrax" became available in English in
1874 in H. von Ziemssen's *Cyclopedia.* Bollinger, experienced with
anthrax in Switzerland, cited industrial outbreaks in Styria and
Waag-Neustadt as evidence that anthrax afflicted "those who work
in industrial establishments where the products of diseased animals are manufactured (especially hides, horsehair, and wool)."
Bollinger concluded that "anthrax in man is, therefore, truly a
disease attaching to certain occupations."[16]

By 1850, anthrax towns had developed in Germany to join the
anthrax districts, long considered the source of the agricultural
disease. Unlike in France and Germany, where the presence of
anthrax districts had stimulated scientific inquiry, physicians and
natural historians in Anglo-American nations had paid little attention to anthrax. While the disease was not unknown, few domestic
episodes of anthrax in livestock or people had been recorded in

Great Britain or the United States before the nineteenth century.[17] No doubt sporadic (and unrecognized) cases of anthrax occurred. Occasional reports of diseases that may have been anthrax in live-stock existed as early as the late 1700s, but British and American agriculturists and physicians did not describe districts plagued with endemic and acute anthrax in grazing livestock as their con-tinental counterparts had, and there were very few reports of the disease in mill workers before the 1850s. Nonetheless, the few cases reported indicated the possibility of infection from contaminated fleeces. As the American physician Pennock wrote, "Observation proves, that the virus causing the malignant pustule, is transmit-ted to man, not only by inoculation from the examination of the dead bodies of these animals, but that of the lightest contact with their hides [or] hair."[18] However, Pennock's was among only a handful of isolated American reports.

British physicians wrote even less about the disease. Perhaps, as Bristol physician William Budd pointed out in 1862, this was because "the malady . . . [had] hitherto escaped general recogni-tion here"; in other words, physicians did not recognize it.[19] How-ever, a wider perusal of the history of wool and worsted produc-tion in Yorkshire mill towns tells a very different story that links the increase in cases of two forms of anthrax, woolsorters' disease and malignant pustule, with changes in the industry. The first case in the British medical literature dated to 1847, and a steady trickle of reports and workers' complaints began to appear in newspapers in the 1850s.[20] In January 1855, a woolsorter at Black Dyke Mills in Queensbury, Yorkshire, died after working with imported alpaca and mohair, and a postmortem examination re-ported "undeniable evidence that the mortality is attributable to the occupation of sorting the mohair and alpaca."[21] Worker agita-tion followed and was renewed again in 1866, when William Cawthra, the foreman woolsorter, and another worker died; let-ters to the *Bradford Observer* urged medical men to investigate.[22] Workers believed imported fleeces to be the most likely cause.

Woolsorters' disease and malignant pustule became a cause célèbre in Yorkshire. Factory workers maintained their own vibrant information networks, and newspaper editors regularly reported cases of anthrax—sometimes in gory detail—for readers outside the factories. In the nineteenth century, in towns with animal-based industries, anthrax was "a subject of great importance," with the "melancholy histories of sudden death" recorded in the local newspapers.[23] It was in these towns that anthrax acquired its identity as an industrial disease.

To understand how this happened, it is important to visit some of these anthrax towns and observe events there in the 1860s and 1870s. Knowledge about anthrax was greatly influenced by local circumstances. Anthrax was right at home in bales of wool, in wool mills and factories, on docks, and even in neighborhoods near these workplaces. Factories became such reliable sources of anthrax spores and vulnerable human populations that scientists in the 1950s would use them as clinical trial sites for new vaccines.[24] Formal investigations into outbreaks of anthrax became more common in the 1870s and 1880s, decades in which scientists and physicians first understood that anthrax had become an established industrial disease, domesticated to factories—indeed, it was one of the earlier recognized occupational diseases. Three major outbreaks documented in the medical and scientific literature occurred in Massachusetts, Glasgow, and Yorkshire between 1868 and 1880. Using bacteriological as well as epidemiological methods, local physicians traced cases of anthrax in mill workers to animal outbreaks in wool-producing regions of the world, marking a new understanding of anthrax's ecology.

MYSTERIOUS AFFLICTIONS: ANTHRAX TOWNS

Walpole, a small city in Massachusetts, is located in the part of New England that was frequently settled by Yorkshire wool workers in the early 1800s. Wool mills had been one of Walpole's major industries for quite some time when, in 1868, workers at the Glover and Wilcomb Hair Factory became ill with a rare and sometimes fatal disease.[25] Silas E. Stone, a Walpole physician, treated and reported on these patients, diagnosing them with malignant pustule (figure 3.1). Stone carried out what we would call

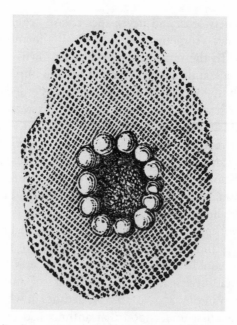

Figure 3.1. Stylized representation of malignant vesicle (anthrax skin lesion). Drawings such as this one helped physicians to differentiate the skin lesions of anthrax from those caused by other diseases. Silas E. Stone, "Malignant Vesicle," *Boston Medical and Surgical Journal* 80 (Feb. 11, 1869): 24. Courtesy of the National Library of Medicine, Bethesda, MD.

a simple epidemiological investigation: he attempted to trace back from the sick patient to determine from whom, or what, the person had acquired the disease.

Stone's cases resonate poignantly, because of the horrible course of the disease and because many of the victims were teenage workers. In most cases, the duration of the disease was a matter of a few days, and the victims were conscious until near their deaths. Of one sixteen-year-old, Sarah R., Stone wrote, "Her mind was perfectly clear. Her answers were correct, and she even smiled when I first spoke to her . . . The distress continued . . . and she died . . . three and a half hours after I first saw her." In the 1860s, Stone saw twenty-five such cases in Walpole factories.[26]

Lacking any evidence of a similar sickness in local people who worked elsewhere, the factory workers implicated the dusty material with which they worked. The people who sickened most often were the workers who received and sorted out the raw materials (horse hair, in this case), and the workers in the early production stages. These workers blamed what they called "Siberian hair"— animal hair from that region, some of which had clearly been taken from dead animals. Many refused to work with suspicious-looking hair for fear of becoming ill, warned other workers, and accused their employers of negligence.[27] In retaliation, employers harassed and fired workers, even subjecting them to lawsuits and pressuring the physicians investigating the outbreak. Ten years later, a *Boston Globe* journalist asked Silas Stone about a report that he had refused bribes to alter workers' death certificates and as a result "was not employed in this disease as formerly." In response, Stone "deftly changed the subject."[28] Because of the powerful public attention to the outbreak, Glover and Wilcomb relocated their factory to Hyde Park, Boston; but even there, woolsorters' disease and malignant pustule followed them. In August 1877, Margaret Donley, a Glover and Wilcomb worker, died "in great agony" from woolsorters' disease. As the *Boston Globe* reported, a second woman became ill and the workers boycotted, declaring that they "might as well be slaughtered outright as die in this manner." The *Globe* journalist traveled to Walpole to interview Silas E. Stone, thus exposing the factory's history with woolsorters' disease at its previous location.[29]

When physicians had interesting and unusual cases, they told their colleagues about them, and Silas Stone was no exception. He read his first paper on the Walpole anthrax outbreak before the Norfolk, Massachusetts, Medical Society on January 8, 1868, and the *Boston Medical and Surgical Journal* published his remarks in February 1868.[30] In preparation for reporting his cases, Stone consulted *Copland's Medical Dictionary* and read the description of malignant pustule presented there by French physician Pierre-François-Olive Rayer.[31] Stone's reading pointed to the fact that most of the published work on anthrax before 1868 available to Americans had been done in France by workers such as Rayer, Toussaint Debrou, Henri Delafond, and Casimir Davaine. All these scientists, physicians, and veterinarians were connected to the thriving Parisian scientific and medical culture of the early to mid-nineteenth century. American physicians and scientists were heavily influenced by Parisian scientific literature, and the more fortunate Americans studied there after medical school to gain valuable experience and knowledge.[32]

These French workers, however, had not written about any association between malignant pustule and woolsorters' disease. Stone focused on his patients suffering from malignant pustule. It is not clear whether he thought woolsorters' disease was related to malignant pustule, but in any case, he was more concerned with singling out the factory and the animal hair as the material sources of the disease outbreak. After describing the clinical courses of his patients, Stone discussed his theories on causation of the disease. "I am inclined to believe that a specific poison is the cause . . . thus far no case of propagation of the disease from one individual to another has been noticed. [But] the leading fact is, that every person was directly or indirectly brought in contact with hair in preparation for upholsterers, or dirt from it, and that in the surrounding population, not so exposed, no cases have been known."[33] Stone reported on five more cases a year later, in January 1869 (again published by the *Boston Medical and Surgical Journal*).[34] By this time, he was working with Richard Manning Hodges, a private physician, adjunct professor of surgery at Harvard Medical School, surgeon and visiting professor of surgery at the Massachusetts General Hospital, and a very well-connected member of the Boston medical establishment.[35] Stone probably called in Hodges (an anatomist and surgeon) because malignant pustule was rare and interesting. Also, surgeons were regularly called in to treat the condition by cutting out the blackened lesions, in hopes of preventing the infection from affecting the whole body. Hodges consulted on one of Stone's cases in 1868 and another in 1869.

Hodges was Stone's link to the rapidly developing scientific investigations of microbes in Europe, and it was Hodges who conducted an early American test of the French investigations. On October 26, 1869, he demonstrated specimens of "bacteridiae" from one of Stone's patients under the microscope to the members of the influential Boston Society for Medical Improvement (Hodges had been a member for fifteen years). Hodges began by describing the paper that Casimir Davaine had presented to the French Academy of Sciences in 1863.[36] Hodges noted that Davaine's "views are generally accepted in France, especially in those parts where malignant pustule prevails." He pointed specifically to three of Davaine's observations: the morphology of the bacteridiae; their lack of spontaneous movement; and that they had to be present in the blood or lymph of the living diseased body.[37] Hodges then compared the organisms he had extracted from the lymph of Stone's factory-worker patient with what Davaine had described.

Hodges called upon the expertise of Jeffries Wyman, another member of the Boston Society for Medical Improvement, a leader in the American scientific community, and a correspondent of Charles Darwin. As Hodges well knew, Wyman's opinion carried a great deal of weight at the local, national, and international levels.[38] Wyman probably knew about Stone's anthrax cases even before Hodges' demonstration, because his brother, physician Morrill Wyman, had conducted an autopsy on one of Stone's patients in June 1868.[39] Jeffries Wyman peered at Hodges' bacteridiae and pronounced them to be the same organisms as those described by Davaine, based on a drawing in Davaine's publication. Wyman probably disappointed Hodges, however, when he declared that Hodges' bacteridiae had nothing "in their form or absence of motion to distinguish them from vibrionides found under a great variety of circumstances." In other words, they could have been ordinary bacterial contaminants; their presence did not prove anything. Wyman "was not inclined to attach to them the specific importance claimed by Davaine"; in other words, Wyman did not believe they had the ability to cause the disease.[40] (Wyman's reasons for these conclusions probably arose from his experiences as a minor figure in the international evolutionary debates of the

1860s, but that is another story.[41]) Even if Wyman didn't think the bacteria played a causal role, Hodges did; and Wyman's confirmation of the bacillus' identity lent weight to Hodges' demonstration and to Stone's case series. For Hodges, Silas Stone, and others who were interested in anthrax as a disease, the identification of the bacillus was paramount. With the validation of Hodges and Wyman, Stone's published case series became a recognized part of the international body of work on anthrax and the development of germ theories.

Information about the Walpole outbreak traveled quickly beyond the United States. Stone and Hodges' *Boston Medical and Surgical Journal* articles were cited by Munich professor Otto von Bollinger in his widely read review article (1874), and the Walpole cases were also included in an article written by Arthur Nichols and published in the *Second Annual Report of the Massachusetts Board of Health*. Nichols' précis of the cases caught the eye of another urban public health official, James B. Russell, the medical officer of health in Glasgow, Scotland. Russell was, by all accounts, extremely well versed in the medical and scientific literature.[42] Almost a decade later, in 1878, when he was confronted with the next significant outbreak of industrial anthrax, Russell remembered the Walpole cases and consulted Nichols's article. In the midst of workers' fears over the mysterious deaths, Nichols's report provided crucial clues for Russell.

The cases in Glasgow in 1878 provided an excellent opportunity to connect woolsorters' disease to malignant pustule, thus suggesting that woolsorter's disease was a type of anthrax. Russell later remembered how he had become aware of an alarming new disease in a local business, the Adelphi Horsehair Factory (figure 3.2): "On 7th March, 1878, rumours reached my ears of a mysterious outbreak of sudden and fatal disease among the workers in a large hair factory in this city. An investigation was immediately set agoing."[43]

Along with this outbreak, Russell was responsible for investigating and mitigating Glasgow's many other public health problems, so his study of the disease in the horsehair workers was

Figure 3.2. The Adelphi Horsehair Factory, Glasgow, 1878. Raw animal hair and fur were stored and washed on the ground floor of this factory, thus potentially spreading Bacillus anthracis spores widely throughout the building. James B. Russell, "On Certain Cases of Sickness and Death Occurring among the Workers in the Adelphi Horsehair Factory, Glasgow," *Rep. Med. Off. Local Gov. Board* 8 (1878–1879): 332. Courtesy of the National Library of Medicine, Bethesda, MD.

largely supervisory and limited to the 1878 outbreak. Nonetheless, Russell's work identified malignant pustule and woolsorters' disease as important new industrial public health problems and fascinating cases for bacteriological investigation. Russell's report of the Glasgow outbreak established bacteriological, along with the usual epidemiological, investigations as the standard for establishing the identity of a disease causing an industrial outbreak. The novelty lay with understanding that this was a newly reported *urban* form of anthrax (which had for so long been an agricultural disease usually seen in rural areas).

Even in the late 1870s, anthrax was still most widely associated with livestock in agricultural areas, so Russell's work was path breaking. Russell recorded how he came to associate the 1878 factory outbreak with "the disease known as anthrax." Here he describes the postmortem examination of one of four women who died in the outbreak:

> The tumour upon her arm was undoubtedly a malignant pustule caused by infection with the animal poison, which conveys the disease known as Anthrax, Charbon, Milzbrand, Splenic Fever, Malignant Pustule, etc., etc. . . . In the blood and fluids of the body of the girl, MD, . . . we actually discovered the contagium of this disease—viz., the *Bacillus Anthracis* . . . In the other cases we had, in the circumstances, no opportunity to discover it; but we can fall back upon the general symptoms . . . from the lips of friends and relatives who witnessed the course of the illness. Having noted those symptoms and compared them with the symptoms in the case of the group in which the *Bacillus Anthracis* was found, and with the account of similar cases put on record from time to time by foreign authors, we obtain conclusive proof of the identity of the disease in the entire group.[44]

In this passage, Russell made an important inference: that the easily identified malignant pustule in one worker was simply a manifestation of the same bacterial infection in other workers—even though they may not have exhibited the skin lesions. Two things linked the workers who had different symptoms: finding the bacilli, and having information about the likely transmission patterns of the disease. Without the latter—the epidemiological information that all of these victims worked in the same place and sickened at about the same time—Russell could not have linked the skin lesion in one victim with the bacteria found in another. In summary, Russell used information from the local workers, the bacteriological analysis from the autopsy of one who had died, and the scientific literature to decide that the various manifestations of woolsorters' disease and malignant pustule were related—that they were anthrax. Russell mostly referred to French investigations, with two important exceptions: Nichols's article in the *Second Annual Report of the Massachusetts Board of Health,* and Robert Koch's work on anthrax in Germany. Nichols's description of "a very complete account of 26 cases which occurred between 1853 and 1870 among the workers in a hair factory in the town of Walpole, in that State" was particularly useful to Russell because "the whole history is so analogous to [that of] . . . Glasgow."[45]

In citing Koch's work on anthrax in 1876, Russell may have been among the first Britons to hear of it. In a lecture delivered in Glasgow in 1876, John Tyndall introduced British audiences to Koch's anthrax investigations prior to their publication.[46] It is not clear whether Russell attended this lecture, but he certainly would have heard of it from his fellow members of the Glasgow Pathological and Clinical Society, the organization that sponsored the lecture.[47] Besides his access to leading British scientific figures such as Tyndall, Russell relied for information on his excellent personal library and the fact that the *Glasgow Medical Journal* (to which he frequently referred) regularly excerpted articles from American and Continental medical journals in the 1870s and 1880s. All these sources helped him to connect the bacteriological and clinical findings and the workers' words and symptoms. Determined to add to the literature, Russell published his report on the anthrax outbreak in the *Report of the Local Government Board* for 1879.

To his disappointment, his medico-scientific treatise, in which he had "endeavoured to give the very broadest expression to the first principles of etiological investigation," was hidden in a supplement to the *Report* behind an article on urban nuisances and sanitation practices.[48] Russell's report provided the first important outline of the argument that linked human cases of anthrax, via bacterial infection, to the materials with which the infected people worked in the hair factory. Sadly for him, and unfortunately for factory workers elsewhere, Russell's important conclusions were little known even by British physicians and public health officers and had little appreciable effect. Certainly, the physicians of Bradford, in Yorkshire, could have benefited greatly from Russell's report. Russell had written, "Since my attention has been directed to those accidents, if they can be called such, in the hair trade, I have had no doubt that 'wool-sorter's disease' is simply internal anthrax. In wool-sorting, almost as thoroughly as in hair-cleaning, trade processes co-operate with the characteristics of the anthrax contagium to produce the most favourable circumstances possible for internal infection."[49]

WOOLSORTERS' DISEASE BECOMES ANTHRAX

While Russell toiled at his report, the number of woolsorters' disease cases spiked in the Bradford worsted mills. John Henry Bell, a prominent Bradford physician, had not read Russell's report (or the account of the Walpole cases) when he began to investigate mysterious illnesses and deaths among some of his patients working in a worsted factory (figure 3.3). (If Bell knew of Russell's report, he did not mention it in his notebook on the woolsorters' disease outbreak, nor did he cite it in his first presentations and publications on the topic.)

Bell did not presuppose the illness to be anthrax. He was dealing with woolsorters' disease, which struck and killed suddenly, often looking like pneumonia or sepsis and not showing the characteristic lesions of malignant pustule. Bell's persistent investigations led to woolsorters' disease being recognized as a form of anthrax in the early 1880s. The Bradford cases established, for the first time, a link recognized by scientists, physicians, and public health policymakers between workers' symptoms and the wool, and between the ill British factory workers and anthrax-afflicted animals in faraway places.

Woolsorters' disease had been an increasing problem in the Bradford district after mill owners (such as Titus Salt) had begun importing raw fleeces and experimenting with ways to use cheaper wools in worsted factories.[50] When workers died of woolsorters' disease, physicians often recorded the cause of death as blood poisoning. Bell was suspicious that this was an occupational problem and that some responsibility devolved on employers to prevent it. Medical officer to the Poor Law Guardians, Bell possessed a political conscience and proved his willingness to act on it. In early 1878, he began his own investigations into the particulars of wool production and processing and into the epidemiology of wool-

Figure 3.3. John Henry Bell. Physician and anthrax investigator at Bradford, Bell encountered anthrax because he was particularly interested in occupational diseases and in improving environmental conditions in Bradford's textile factories. Courtesy Wellcome Library, London.

sorters' disease. Simultaneously, he used the local press and the legal instrument of inquests to indict working conditions as the cause of the disease.[51] Bell was increasingly able to obtain access to sick woolsorters' bodies, before and after death, because concerned families gave permission for autopsies.

These cases and the woolsorters themselves were the source of much of Bell's early information on the disease.[52] Bell made a point of asking both management and workers their opinions, which he recorded in his notebook. At first, he approached the disease as a sanitary problem: "Are they [the sheep] washed before shearing?"; "Salts [Saltaire factory workers]: Mohair washed on back; Alpaca not washed; Higgins [foreman]: thinks not, or wouldn't be so dusty; Salts men [workers]: not washed."[53] The workers pointed to the fleeces they sorted as the source of the disease and tried to avoid working with "bad" bales of fleece.[54] They suspected that imported fleeces—the alpaca and the Van mohair from Asia Minor—had some kind of dusty poison in them. The power of their assertions, and their importance to Bell's investigations, can be better understood in the context of the workers' overall situation in the mid-nineteenth century. Since 1818, Bradford woolsorters had formed a mutual relief society that virtually controlled the local job market in their trade.[55] The Bradford Woolsorters' Society regulated the specialty workforce, controlling employment and ensuring good relationships with employers. It also provided benefits when workers were ill, contingent on a physician's examination or steward's certificate. The expense of a physician was necessary to avoid the fines levied by the society (for missing work); a worker paid either the physician or the fine.[56] Therefore, physicians were called in on many cases of woolsorters' disease that otherwise might have remained unknown and unreported. Woolsorters viewed physicians as allies in their quest to hold employers accountable for occupational disease, and the workers became important partners in the scientific investigations.

Bell wanted to make contributions to scientific knowledge about industrial diseases, and the woolsorters' disease outbreak was his chance to conduct an investigation and publicize his findings. He began with the workers' assertions—that the imported wool was poisonous—when he presented his first paper on woolsorters' disease to the Bradford Medico-Chirurgical Society.[57] But what was the exact nature of the "poison" causing the disease? How could he confidently define it and understand its actions on the human body? A turning point came for Bell when he attended the February 1878 meeting of another medico-chirurgical society, in Leeds, an industrial city thirty miles distant in which he had trained as a young physician. After a conversation with another physician at the meeting, John Eddison, Bell realized that woolsorters' disease could be related to anthrax. Eddison, who had recently completed a period of study in French laboratories and clinics, told Bell of experimental work there on splenic fever (or anthrax) in animals and the linking of a specific bacterium to the disease.[58] About two-thirds of the way through his notebook, Bell wrote the word "anthrax" for the first time. He began to experiment with bloody imported wool—to follow the same procedure Koch, Pasteur, and others had used before him to demonstrate anthrax in animals. At first, Bell's experiments failed; but late in 1879, he successfully injected blood from a patient who died of woolsorters' disease into experimental animals, which then also died.[59] He concluded that woolsorters' disease was probably a type of anthrax, or splenic fever. The hardy spores, described two years earlier by Robert Koch, clung to the hair and hides of ill animals and transmitted the germ of the disease after gaining access to the body of a living animal. Bell had previously approached this industrial ailment as a sanitarian or epidemiologist; now he switched to bacteriology. This switch represented Bell's entrance into the transnational community of anthrax investigators. The publications of his results in the *British Medical Journal,* and his role in developing anthrax-prevention guidelines, established him as a "pioneer in occupational anthrax."[60] Indeed, Bradford temporarily became the location of "a leading edge of medical research and a testing ground for the emergent germ theory of disease."[61]

Painstaking bacteriological and epidemiological investigation transformed woolsorters' disease into anthrax in 1881 in Bradford. Bell's work assumed national importance when, in 1880, the Home Office ordered John Spear (a local government board inspector) to begin an investigation. (Spear almost paid with his life after cutting his hand during an autopsy of a woolsorters' disease victim and coming down with malignant pustule.)[62] Spear engaged William S. Greenfield, professor-superintendent of the Brown Institution, to conduct the laboratory experiments on bodily fluids and tissues from victims of the disease. Greenfield used his findings as the basis for his 1880–1881 Brown Lectures, in which he declared the anthrax bacillus to be the definitive cause of woolsorters' disease.[63] Greenfield had followed the bacterial traces of the disease from blood-stained mohair through the bodies of workers and experimental animals, and in so doing, he created a persuasive body of evidence that would eventually influence public health policies directed at both animals and people. Greenfield's data provided widely respected evidence that anthrax had become an industrial disease.

Dramatic events that took place in May of 1880 provided even more direct evidence for the link between anthrax in animals and woolsorters' disease. In the Bingley district of Yorkshire, a farmer named Dunlop kept cows and sheep on lots near Watmuff's Mill. In April and June, three Watmuff workers had become ill, and the woolsorters refused to continue working with Van mohair. The mohair was washed and processed without being sorted, and the wash water was drained onto the lots housing Dunlop's animals. Three cows and two sheep promptly died, and both veterinarians and physicians conducted postmortems on the animals. John Bell attended one of the postmortems, and he sent blood and tissue samples to Greenfield, who confirmed the presence of anthrax bacilli. In Spear's 1881 *Report to the Local Government Board*, he wrote, "This disease amongst the cattle, in short, was demonstrated to be identical with that amongst the woolsorters, and it seems certain that the infection was obtained from the same source."[64] Spear had echoed workers' claims by focusing almost immediately on the mohair imported from Asia Minor; now the disease had come full circle, infecting animals thousands of miles away from those in which it had originated. "The evidence," Spear concluded, "may be said to be complete."[65]

Complete evidence notwithstanding, Spear took one more interesting step to investigate anthrax. Working through the Bradford manufacturers, he solicited information from wool-purchasing agents living in Constantinople and Peru about diseases in goat herders and wool workers. They reported that anthrax-like syndromes were well known and seemed to be tolerated without alarm by the local people. According to an agent named Gatherell in Constantinople, Turkish workers were very familiar with a disease of mohair woolsorters, *dallack*, which caused malignant pustules. Although the agent wrote that his account undoubtedly "contained some nonsense," he described a flourishing medical specialty of *dallackjis,* or dallack doctors, who concentrated solely on this disease. The workers had great confidence in the dallackjis, who lanced and cauterized the pustules, applied herbal poultices, and prescribed diets of bread and caviar. This confidence meant that workers were not reluctant to handle Van mohair, believing that the dallackjis would cure them if they contracted the common infection. The word *dallack* meant "spleen," and the dallackjis informed the agent that the disease was so named "because that is the organ which is diseased in the animal from which the infection comes." Dallackjis also recognized different types of dallack depending on which species of animal it had come from. Gatherell and one Dr. Patterson, a Scottish physician posted at the English Hospital in Constantinople, acknowledged that few workers there ever died of dallack.[66] Long domesticated in the ancient cradle of goat and sheep herding, anthrax had a place in the medical classification system and the treatment framework, and a group of professional Turkish healers devoted to it.

SAME BACILLI, DIFFERENT DISEASES

With the increasing international wool and hair trade, anthrax had moved into new locations whose physicians had little experience with it. The different medical understandings of anthrax, from woolsorters' disease in Britain to dallack in Turkey, provided hints of the bacillus' and the disease's history in a given location. In Turkey, woolsorters were said to work in the open air, which helped them to avoid inhaling many spores. Dallack seemed to be a less deadly and more common infection (mostly in the form of malignant pustules), with those sickened being tended by medical specialists devoted to just this disease. But the same spores in Britain (if indeed they were "Turkish" spores) caused woolsorters' disease, death, and fear in factory workers. Were these the same spores, and if so, why did they act so differently in different places? In part, human agency affected how deadly the spores were. For example, if the British woolsorters had worked outdoors instead

of on the first floor of a factory building, the incidence of woolsorters' disease might have been lower. The historical evidence
provides the clear picture of a disease that, although caused by the
same bacilli, operated very differently in different places. It is useful at this point to put the historical evidence in conversation with
other forms of evidence, specifically what we know about the bacillus' biological attributes.

Doing so opens up new questions and avenues for future research about the history of microorganisms (such as *B. anthracis*)
that are associated with diseases. To understand how this might
work, a brief review of some aspects of *B. anthracis'* life cycle (as
we currently understand it) is in order. *B. anthracis* spores live in
the soil for long periods, and they are likely to have been resident
for the longest period in the area where grazing animals were first
domesticated: the Fertile Crescent (which includes the region of
Turkey around Lake Van). Once a grazing animal picked up the
spores (usually by eating them), the bacillus could not complete
its life cycle unless it killed the animal, thus releasing more spores
into the environment around the carcass. Therefore, the next generation of bacilli traveled only as far as the animal host traveled
before it died. Like Darwin's finches in the Galapagos, *Bacillus
anthracis* evolved over time in a very limited physical area, adapt-
ing to a particular environment.

Enter the historical record, which suggests that human agency
transported these bacilli in animal hair or wool to completely new
environments in the nineteenth century. Were the spores the same,
and if so, what did their relocation to a new environment mean in
terms of what constituted "anthrax"? It seems useful, at this point,
to look for genetic changes in the bacilli over time. In 2007, a
group of geneticists traced the 150-year-old footprints of *Bacillus
anthracis* in the global wool trade. Like John Bell, Matthew Van
Ert and his colleagues sought to understand the ecology and distribution of anthrax. They sampled over 1,000 strains from around
the world and compared genetic markers to see how, and how
often, the bacilli had experienced mutations and other genetic
changes. They could also see which strains in one place were ge-

netically related to strains in another, thus suggesting that some bacilli had been moved to new places in the past. By tracing the strains to their geographic "donor" regions (much as Bell had done with his epidemiological inquiries), the geneticists were able to begin piecing together a pattern of *B. anthracis'* movements over the past few hundred years.[67]

Van Ert and his colleagues wrote that certain strains of *Bacillus anthracis* had traveled the world, experiencing "great reproductive success . . . and considerable long-distance dispersal." Related strains of the bacillus could be found in places quite distant from each other: China, Eurasia, South America, North America, and Europe. The older, rarer strains lived in much more narrowly focused geographic areas in the Middle East and on the continent of Africa and had not moved very much. Van Ert and his colleagues remarked that without the tremendous expansion of *Bacillus anthracis* around the world, anthrax would be a "highly restricted and rare disease" in the early twenty-first century, and thus of little interest or concern to scientists and physicians today.[68] This dispersal of *B. anthracis* around the world also affected its evolution. What had helped the world-traveling strains to outstrip the older strains, thus becoming the dominant family of *B. anthracis* and keeping the disease from being a rarity?

Van Ert's group acknowledged that several theories could help to answer this question and explain their results. Perhaps the local environments or a paucity of animals to infect had restricted the older, rarer strains' growth and expansion, leaving the field wide open for the dominant family of *B. anthracis*. This could lead to possible genetic differences in the strains as they evolved over time and adapted better to their environments. Van Ert and his colleagues decided, however, that strictly biological theories of adaptation did not explain the pattern of their data. Instead, they concluded that their results supported a pattern of "human activities in commerce and industrialization . . . impacting the global population structure of *B. anthracis.*"[69] Within the past two hundred years, a "cosmopolitan assortment" of *B. anthracis* strains had been imported into the United States and Western Europe.

These strains had originated in regions from which large amounts of wool, hair, and hides had been imported in the nineteenth and early twentieth centuries.[70] Thus, the geneticists concluded, human agency had at least contributed to, and probably caused, a major event in *Bacillus anthracis'* history and evolution: the tremendous expansion of certain strains of *B. anthracis* and their circulation around the globe in the nineteenth and early twentieth centuries. As Van Ert's article concluded, "human activities have dramatically influenced its [*B. anthracis'*] current distribution and occurrence."[71] In other words, the genetic evidence circa 2007 and the historical evidence circa 1880 together give us a richer picture of anthrax than either type of evidence would have done alone.

That picture tells us that, although the bacillus remained genetically the same, the disease could look very different from place to place. In other words, the bacillus did not determine the shape of the disease, even though the bacillus was viewed as the cause of the disease. Spear's report and the Van Ert group's genetic data both asserted that particular strains of *Bacillus anthracis* that had originated in eastern Turkey had traveled to Britain and caused woolsorters' disease and malignant pustule. The genetic data circa 2007 showed that the transplanted strains of *B. anthracis* in Britain were still very closely related to bacilli remaining in Turkey. (In other words, they had not yet had time to evolve a great deal.) Indeed, they were so close in genetic makeup that Van Ert and his colleagues considered them to be the same strain in 2007. Yet the historical record very clearly distinguished between dallack in Turkey (a common, mild to moderate disease) and woolsorters' disease in Britain (an uncommon, but very lethal disease), even though they were caused by genetically identical bacilli. Many reasons probably contributed to these very different disease experiences: what people believed about anthrax, how they handled the wool, and how strong their immunity would have been, just to name a few. Answering this question requires more research, but for our purposes, the important point is this: anthrax the disease was not equivalent to *Bacillus anthracis*. The same bacillus caused what looked like different diseases. The identification of the bacillus as a common factor in all these disease types did link the manifestations of woolsorters' disease, malignant pustule, and dallack, using bacteriology to connect the global wool trade to the mysterious disease in British workers. This opened the door to creating and enforcing public health responses to the threat of anthrax, although the best intentions did not always—or even usually—translate into rapid and successful control of the disease.

CONTROLLING INDUSTRIAL ANTHRAX

Agricultural anthrax had long been combated in Germany and France by a combination of local and state-administered public health measures, and this would be the pattern for controlling industrial anthrax as well. Woolsorters' disease carried a far higher fatality rate and was a special public health focus. Local woolsorters' societies, trade unions, wool and worsted mill owners, and the anthrax towns' public health officials and chambers of commerce all had a hand in determining public health responses to industrial anthrax. In their efforts to prevent or at least control workers' exposures, these groups faced both political and technical problems. Predictably, manufacturers' fears of losing money led them to oppose stringent regulations and bans on importing the cheapest (even though the most dangerous) raw materials. In some areas, regulations were created but not enforced, for similar reasons. Officials also faced choices in the types of actions that should be taken. Could foreign sources of the disease be somehow cleansed of their impurities through a new technology that killed the spores but preserved the fleeces? Could factories, sewage systems, and adjacent fields be cleaned and disinfected—a traditional sanitary solution? Or perhaps it would be possible to bolster the strength of workers' resistance to the disease; physicians had long debated why some workers sickened and others did not, and the answer seemed to lie in the mysteries of the human immune system. Physicians and bacteriologists continued to search for treatments, emboldened by the development of diphtheria antitoxin, tuberculin, and other potential cures for bacterial diseases in the mid-1890s.

Public health officials, physicians, and others interested in controlling anthrax explored technological, sanitary, and medical solutions to the problem of industrial anthrax (each costly in time and money).[72] Different nations settled on different overall strategies that reflected their political and scientific resources and traditions. German provinces relied on a system of sanitary policing but continued to experience industrial outbreaks well into the twentieth century. The French, following the Pastorians' vaccination success in 1881, relied on vaccines and other medical preventives. So did Italy, home of the first widely used anthrax antiserum, developed by Achille Sclavo in the 1890s. While the United States relied on its geographic isolation to decrease the danger of importing infected fleeces, Great Britain chose the technological fix of building a disinfection plant at Liverpool to treat the most dangerous imported materials.

In Britain, the political power of manufacturers steered the nation toward the technological solution of finding a way to disinfect imported materials rather than banning them outright. In an initial dearth of state concern and activity, local groups in Bradford began in 1884 to determine cooperative strategies for controlling anthrax in the hard-hit district. This followed from John Bell's insistence on publicizing the results of coroner's inquiries into deaths among woolsorters. Bell threw down the gauntlet before mill owners by certifying that a woolsorter's death in 1880 resulted from his employer's neglect in not properly washing or disinfecting imported mohair. Local woolsorters, their families and friends, and the activist publisher of the *Bradford Observer,* William Byles, all agitated for reforms to protect the workers. The Bradford Medico-Chirurgical Society appointed a commission from among their members to look into the anthrax question. The members included both bacteriological proselytizers (such as John Bell) and skeptics (such as Edward Tibbets, a local medical practitioner). The commission took testimony, read the medical literature (Davaine, Bollinger, Cohn, Koch, Burdon Sanderson, Greenfield, and others), and even conducted a few experiments. In 1882, they published their *Report of the Commission on Woool-*

sorters' [sic] *Diseases.* The report's controversial character may be divined by its limited print run and the "Printed for Private Circulation Only" stipulation on the cover.[73] The experiments had not proved to be as definitive as William Greenfield's conclusions had been, reinforcing the skepticism that *Bacillus anthracis* (and not dust or other septic poisons) caused the disease. This distinction mattered, since the legal grounds for forcing employers to alter their importation and wool-handling processes rested on proving a bacteriological etiology and associated preventive measures for the disease.[74] Unable to be unanimously decisive, the physicians of Bradford could not lead the effort to shape public health responses to the problem of woolsorters' disease.

Instead, in 1884 a consortium of business interests, physicians, and local officials developed the Bradford Rules, the first British public health response to industrial anthrax. The rules specified the nations of origin from which bales of imported wool had to be classified as dangerous, a list headed by East Indian cashmere and fleeces, Van mohair, and Egyptian wools. These fleeces were to be sorted apart from others, and wool-sorting rooms needed to be ventilated with down-draft fans. Most employers complied with the guidelines, and the few that did not were vulnerable to coroner's juries that could hold them criminally negligent. These and other guidelines could not be legally enforced, but the woolsorters, their families, and their physicians kept the public spotlight trained on employers. Physician Frederick Eurich clipped a poem from an issue of the widely circulated newspaper the *Yorkshireman:*

I should, were I of hasty turn,
With indignation boil
To think of those who fortunes earn
By letting workmen toil
Amid the deadly Van Mohair
When at a little cost and care
Their lives might shielded be.[75]

The physician John Bell also used the newspapers, especially the *Bradford Observer,* to urge the workers to protest: "Will you quietly go on being poisoned from day to day, struck down in a few hours with a horrible pestilence, rather than speak out like men, and refuse to touch the abominable stuff till it has been properly cleansed by those whose duty it is to cleanse it."[76] Memories of the sometimes violent Chartist worker uprisings of the 1840s doubtless added weight to these public expressions of woolsorters' dissatisfaction. In the years leading up to the outbreak in Bradford, woolsorters' societies reorganized, town councils and employers came to broader political understandings, and a city medical officer of health was appointed for the first time. Thus, the anthrax outbreak of 1878–1882 took place in a dynamic political and social environment. The old political alliances from mid-century were breaking down, and a working-class movement for political independence, tied into the developing public health framework, was beginning.[77] Politics and woolsorters' disease were tied together in Bradford, and the district's position as the nation's leader in worsted production meant that local political solutions became national policy.

The Bradford Rules became the basis for the Home Office's first attempt at national regulation of industrial anthrax, the 1895–1896 Special Rules for woolsorting; but political infighting hamstrung the Special Rules from the start, and dissatisfied workers and physicians continued to press for stricter guidelines.[78] The energetic physician and factory inspector Thomas Morison Legge used his political position to reinstate interest in occupational anthrax in the early 1900s.[79] In 1905, Bradford again took the lead in appointing a consortium of physicians, employers' representatives, and trade union officials to the Anthrax Investigation Board. The board hired physician Frederick Eurich to investigate the bacteriological dangers of specific classes of wool and hair. Eurich determined that the traces of blood on the fleeces, rather than the wool and hair itself, harbored the bacilli, and that disinfection procedures should be structured accordingly.[80]

The intransigence of anthrax—its persistence in the environment, the inability to control the importation of fresh spores, and the potential lethality of the disease—must have been quite frustrating for physicians such as Legge and Eurich.[81] So much was known about the biology and ecology of anthrax and the locations where it could be expected, yet it could not be eradicated or controlled. In 1913, G. Elmhirst Duckering spearheaded the effort to find ways to kill the anthrax spores without ruining the fleeces. Duckering's group conducted more than one hundred experiments to finally perfect the process, which involved washing the wool in 2.5 percent formalin. Duckering's success, and the confidence of his report, heartened industrial leaders, particularly given an alarming increase in anthrax cases during World War I.[82] If fleeces could be successfully disinfected, then the mill owners' opposition to import bans would no longer be a barrier to protecting workers' health.

One hurdle remained: industrialists' aversion to regulations requiring their industry to foot the bill for mandatory disinfection. Under the Anthrax Prevention Act of 1919, however, and in a political climate dominated by Labour Party reformers, they ended up doing just that, since the new disinfection plant erected at Liverpool was financed by tariffs on imported fleeces. Even so, the disinfection station ran £37,000 in the red in its first year, coinciding with a depression in the East Indian wool trade and in wool textiles production.[83] The disinfection plant has been credited with contributing to the steady decrease in industrial anthrax cases between 1921 and the 1960s, when it closed. Historians have pointed out, however, that the plant's limited capacity meant that not all dangerous materials passed through it (only Egyptian fleeces and Indian goat hair, at first), and that the risk to workers simply passed from the mills to the disinfecting plant, where sixteen workers contracted anthrax during its forty years of operation.[84] Nonetheless, Britain was the first major manufacturing nation that attempted to centralize importation and disinfect imported raw materials to prevent anthrax. The numbers of cases did decrease after the plant began operating—either because of its existence or because much of the cheap, dangerous raw materials were diverted to other nations without disinfection laws.[85]

Germany at first adopted a very different approach in the 1880s because of longstanding political traditions and an emphasis on bristle and horsehair processing, animal products not easily disinfected en masse. State regulations to address agricultural anthrax had been in place since the late 1860s, stipulating proper handling of animal carcasses, disinfection of premises, discarding of infected materials, and worker-protection recommendations. Unfortunately, as Munich professor (and physician) Otto von Bollinger complained, employers often ignored or circumvented the regulations.[86] Medical men also did not agree on the necessity for such stringent regulations. In 1869, the major German textbook of veterinary public health procedures (*Staatstierheilkunde,* or state animal medicine) argued that regulation of products such as wool should not be required, since the author thought anthrax was not truly contagious; educating the workers to maintain a high level of personal hygiene would suffice.[87] In response, Bollinger warned that without "exercising the strictest police control," the disease would continue to plague German agriculture and industry.[88] Although veterinarians and physicians were not required to notify state authorities of anthrax cases, estimates indicate about ninety human cases per year in the last quarter of the nineteenth century, with a 10 to 30 percent mortality rate. Most of the fatal cases occurred in the horsehair and brush-making mills and in the tanneries. At one factory, the German Empire Horsehair Factory at Schleswig, twenty-five of the thirty-five employees became ill with anthrax in an eight-year period (1873–1881), and almost half of them died.[89] German factory towns now joined agricultural anthrax districts as places in which the bacillus had become domiciled. Anthrax continued to be an important presence in Germany well into the mid-twentieth century.[90]

In the United States, cases of anthrax continued, seemingly unabated, among hide, wool, and hair workers throughout the early decades of the twentieth century. Britain's tariffs meant that more of the cheaper, more dangerous raw materials made their way to American ports. With a decentralized and relatively unregulated importation system, a central disinfection plant (or plants) was impractical. Instead, in 1916 the federal government developed regulations (over the objections of the affected industries) that required raw animal products from anthrax-endemic regions to be disinfected prior to entry into the country. These restrictions worked to some degree, reducing the flow of diversions from the British market and curtailing imports from other areas such as South Africa (much to the dismay of foreign livestock producers).[91] Thus American isolationism, so prominent in the years leading up to World War I, lay at the heart of the strategy for anthrax prevention. But this did not prevent American scientists, especially federal veterinarians at the Bureau of Animal Industry, from attempting to develop medical solutions to the problem, as we will see. Nonetheless, the small numbers of anthrax cases and domestic livestock raisers' interest in maintaining importation control meant that the United States could adopt a limited anti-anthrax policy based on education and better hygiene and still not experience many cases of the disease.

France adopted another strategy, a predominantly medical model of vaccination and serum treatments that became national policy due to the considerable marketing talents of French scientists. As we have seen, after the Pastorians' dramatic public vaccination demonstration on the sheep at Pouilly-le-Fort in 1881, the development and use of prophylaxis was a matter of national pride. Restricted by French law from patenting a veterinary biological agent such as a vaccine, the Pastorians kept critical details to themselves to maintain control over the production and mass diffusion of the anthrax vaccine throughout the 1880s and 1890s.[92] Their product was widely used in France in the first few years after Pouilly-le-Fort. About two million animals were vaccinated in France between 1882 and 1887.[93] In addition, Pasteur and his colleague Chamberland oversaw the system by which laboratories to produce Pastorian vaccine, "guaranteed in quality," were established around the world.[94] Local companies entered into formal contracts with the Pastorians, who supplied the vaccine and the know-how to set up the local institutes.[95] France remained the center of the medicalized strategy to vaccinate the nation's animals in order to protect its human citizens—a strategy that might seem to emphasize agricultural anthrax over industrial. In regions such as Saint-Denis, home to horsehair-processing and leather-dressing factories, industrial anthrax was still a problem that caused deaths, and a safe and effective vaccine for humans was decades in the future.[96] However, public health officials and physicians considered infected animal populations to be important sources of recurrent infection in people. Combating anthrax in animals was a necessary early step in eradicating the disease from human populations as well.[97] The French, with the combined chemical and commercial savvy of the Pastorians, led the world in the medicalized solution of mass vaccination against anthrax, at least until World War I disrupted vaccine distribution. Then many nations, most notably South Africa, which housed talented scientists at the Onderstepoort Veterinary Research Institute, sponsored efforts to develop their own supply of vaccine. In the footsteps of the Pastorians, the Onderstepoortians would also hold a monopoly on the supply of anthrax vaccine within South Africa.[98]

The French strategy represented the beginning of an important trend away from the cornerstone of anthrax control, which had been to reduce contamination in the external environment. The external environment included factory rooms, with contaminated soil, dust, horsehair, wool, bone meal, and other animal products. Combating anthrax meant cleansing these parts of the human and domesticated animal environment to remove the spores, the source of infection. But a new strategy was required as anthrax became domesticated in more locations. Attempts to disinfect the external environment were failing to halt outbreaks; the *Bacillus'* spores were too persistent. Instead, public health officials in most industrialized nations slowly turned to controlling anthrax by altering the internal environment of its victims, both human and animal. Increasingly after the turn of the twentieth century, anthrax investigators and public health officials began to look for ways to fortify potential victims against the infection. With the development of vaccines and a growing understanding of the science of immunology, scientists and physicians focused on creating anti-anthrax prophylactics and treatments that could sidestep the technical and political problems of building disinfection plants and controlling imports of hair, hides, and wool. Historians have noted this shift in strategy for other infectious diseases and for chronic diseases in the mid-twentieth century, but anthrax once again was an early test case.[99] Scientists and physicians had intimate knowledge of its life cycle and ecology, and knew well how to manipulate it in the laboratory. The bacillus had become domiciled in the soil, laboratories, and factories, and the bodies of people and animals. Anthrax had always been a disease of place; now anthrax investigators transferred this view to bodies as domiciles for anthrax.

ALTERING THE INTERNAL ENVIRONMENT TO CONTROL ANTHRAX

Altering the internal environment of anthrax victims' bodies meant developing a series of vaccines, sera, and antimicrobial compounds that could either stimulate the victim's immune system, kill the bacteria, or do some combination of both. Around 1900, with the Pastorians effectively controlling the business of animal vaccination, other investigators used the young science of immunology to develop other products and procedures. Investigators tried normal beef serum (from unvaccinated animals), extracts of *Bacillus pyocyaneus,* and horse serum. None provided adequate protection in test animals, and some caused severe adverse reactions. In 1895, Emile Marchoux, a Pastorian and director of the Laboratoire de Saint-Louis du Sénégal, vaccinated sheep by the Pasteur method and then began to hyperimmunize them over a period of weeks with virulent anthrax cultures. Marchoux produced an immune sheep serum that protected experimental animals when it was administered up to twenty-four hours after injection of virulent anthrax cultures.[100] Marchoux's method was impractical, and he was hampered by his provincial location, but an Italian scientist working independently, Achille Sclavo, also developed a workable anti-anthrax serum in 1895, using donkeys instead of sheep as his living serum factories. Sclavo's work was born of necessity: Italy faced a serious anthrax problem, with a preponderance of agricultural cases in the south and more industrial cases in the north. Between 1890 and 1904, 36,436 cases of anthrax were registered among tanners, brushmakers, and wool workers, with 7,308 deaths.[101]

The serum was so promising that physicians from around the world made pilgrimages to Siena to learn more about it. "Sclavo's serum" appeared as a notation in John Bell's field notebook in 1904, and it was inspiring great excitement among anthrax investigators in Britain. Thomas Legge visited Sclavo's laboratory and brought back glowing reports of its efficacy as well as a stock of the serum.[102] At his home outside Siena, Sclavo founded the Istituto Sieroterapico e Vaccinogeno Toscano to develop, manufacture, and sell the serum (the proceeds of which supported the institute and Sclavo). Like the Pastorians, Sclavo and his colleagues had a monopoly on the production of sera and vaccines in Italy. Similar to the situation in France, Italy adopted the Sclavo serum as the centerpiece to its medicalized anti-anthrax strategy.

Scientists debated the serum's method of action. Did it alter the immune system response, or have bacteriocidal properties, or some of both? It certainly contained agglutinins, substances that caused cells to clump together (thus potentially destroying the bacilli), but these properties could vary widely between patients and strains.[103] The serum likely contained antitoxin properties, but this could not be known for certain since scientists had not yet isolated toxins produced by *Bacillus anthracis*. Regardless of how it worked, physicians around Europe published many cases over the next forty years of anthrax victims treated successfully with the serum. Even internal cases of anthrax septicemia, thought to be 100 percent fatal, had been efficiently cured with intravenous injections of the serum.[104] Of course, it had drawbacks: the serum was expensive (U.S.$10–$30 circa 1920); it had a limited shelf life (Sclavo's laboratory claimed two years, but others reported that it did not last that long); and large doses had to be injected into patients (up to 100 cc at a time). Such large injections caused pain and could cause allergic reactions.[105] Nonetheless, serum remained a mainstay of treatment for malignant pustule and inhalational anthrax, especially in Italy and South America.

In the United States in 1915, veterinarian Adolph Eichhorn at the Bureau of Animal Industry oversaw the development of an even more potent serum (using a different vaccination method than the Pastorians). Eichhorn's serum was used on animals and people alike in the United States, with the bureau distributing it widely to hospitals and veterinarians. By 1921, however, this product had also become commercialized, and the bureau yielded its production to private laboratories.[106] Hyperimmune sera had become easily manufactured, easily obtained, and widely used. Along with the early vaccines, sera represented the first effective way to alter the internal environment of anthrax victims.

In 1909, salvarsan, an arsenical compound, was developed to treat syphilis, and physicians became keen to try this on anthrax cases as well. Some felt that it worked as well as Sclavo's serum, and it was cheaper and easier to keep in stock (a concern when immediate treatment was recommended in factories and provincial areas far from a vaccine-producing laboratory). Adrianus Pijpers, pathologist to the Pretoria Hospital in South Africa, asserted that not one person treated had died in a series of forty cases he had treated with salvarsan. Doses of 0.6 to 0.9 grams intravenously were common, and "the worse the patient appears to be, the bigger the dose." Pijpers asserted that, although the "theoretical foundation" for this treatment "may be lacking, the salvarsan treatment of anthrax has never failed us yet."[107]

In the United States, an inability to control anthrax in the external environment stimulated the Bureau of Animal Industry and private companies to try to develop agents that could alter victims' internal environments prior to infection by using vaccines. Anthrax outbreaks in the United States were widespread, occurring in mills as well as in rural areas. In 1921, more wool on the global market that was likely to contain *B. anthracis* spores got diverted to the United States, as the opening of the disinfecting station at Liverpool caused British tariffs on imported wools to rise. The manufacturing states of New York, Pennsylvania, and Massachusetts experienced the highest rates of industrial anthrax, while cattle-producing and importing regions (Louisiana, Texas, the Dakotas, and California) were home to most cases of agricultural anthrax. Veterinarians in these regions depended heavily on vaccines to combat the disease. By the late 1930s, Louisiana veterinarian Louis Leonpacher and his young associate vaccinated up to sixty thousand animals annually.[108] By this time, however, most American anthrax cases in animals and humans were treated with an Eichhorn-type serum. Right up until the development of antimicrobials such as sulfa compounds, serum treatment was the predominant method of treating anthrax cases in the United States.[109] American scientists also worked to create a system of prevention that combined a German-developed serum with various vaccines, but this regimen was very complex. Physicians and veterinarians needed an effective, safe, inexpensive single-dose vaccine with a long shelf life that could be easily kept and used in factory and pastoral conditions.

Since the Pastorians' success, scientists in many nations had been experimenting with various vaccine types. In the 1930s, just before the development of antibiotics, South African veterinary scientists tried a radical new approach to developing their own anthrax vaccine. At the Onderstepoort Veterinary Research Institute just north of Pretoria, South Africa, veterinarian Max Sterne isolated what became known as the Sterne strain of anthrax in the early 1940s. Sterne had approached the problem of how to create a safe and effective vaccine differently from his predecessors. Anthrax vaccines had previously been created from virulent strains, under the assumption that a strain had to be virulent to produce immunity. Sterne chose to follow the lead of thirty-year-old German studies that showed that a few naturally occurring strains could produce immunity in experimental animals without ever producing disease.[110] Although Sterne and his colleagues were unaware of how exactly the avirulent strains were different, we know now that they lacked a protective capsule, which made them vulnerable to destruction by the host animal's immune system. They were easily identified because the colonies they formed on artificial culture media in the laboratory looked markedly different to the naked eye. The bacilli that lacked the exterior protective capsule formed ragged-edged colonies, while those that retained their capsules (and their virulence) formed smooth-edged colonies. The crucial task was to culture the desired colony types under laboratory conditions. Sterne's colleague J. H. Mason developed a semisolid culture tube that made it possible to undertake the painstaking process of growing new strains and testing them for virulence and immune protection in experimental animals.[111]

Sterne's breakthrough depended heavily on an important assumption: that anthrax developed differently inside the body of an animal than it did in the "hostile" environment of laboratory cultures.[112] Once Mason and Sterne had created culture conditions that replicated the conditions inside an animal's body, anthrax responded by growing into the colony types that the scientists needed to propagate the avirulent vaccine strains. Sterne, Mason, and their colleagues had managed to grow bacilli and to manipulate the physical structure of *Bacillus anthracis* by altering the environment in which it was raised. The avirulent vaccine strain caused fewer reactions and disease in vaccinated animals than the Pastorian method had. Yet it still proved effective at protecting vaccinated animals challenged with anthrax infection.[113] The South African government wasted no time in adopting the Sterne vaccine for use in livestock. Within two years of its development in 1936, the vaccine was being given annually to over half the animals in the country. In historian Daniel Gilfoyle's words, this was "a major state intervention in rural society," especially in the homelands of black South Africans, where Onderstepoort scientists first tested their new vaccines on cattle under a state-issued compulsory vaccination order.[114]

By thinking of anthrax as an environmental disease, and by thinking of bodies as environments, the Onderstepoort veterinary scientists had figured out a way around problems that had stumped the Pastorians and everyone else for almost fifty years. They also developed new techniques for manipulating the bacilli, maintaining what was essentially an artificial breeding program that greatly altered populations of bacilli in predictable ways. This work heralded a new era of strategies for controlling anthrax. The Onderstepoortians had propagated a strain of anthrax that lacked a certain part of the *Bacillus anthracis'* genetic apparatus. One plasmid, a piece of circular DNA, was missing from the Sterne strain bacilli. This plasmid, later named pXo2, encoded for the production of the organism's protective capsule. Sterne had called it an avirulent strain, but this wasn't quite technically true. This strain still retained its ability to manufacture lethal toxins in an animal's body (a property encoded on another plasmid, pXoI, still in the Sterne strain). In other words, these bacilli had retained some of their killing power. Had he known this, Sterne might not have believed his favored strain to be reliably harmless. Nonetheless, in extensive testing in laboratories and in the fields where black South African herders kept their cattle, almost no cases of infection could be attributed to the Sterne strain. As Peter Turnbull has described, Sterne received little scientific or economic return, but his strain was "freely handed out on request in the spirit of generosity" to scientists in other nations who produced vaccines from it.[115]

Laboratory-manipulated strains of anthrax that lack the pXo2 plasmid remain the basis for many of today's anti-anthrax vaccines. By the 1950s, the Italian Sclavo institute was selling its vaccine abroad to other nations (including the United States).[116] In Britain, a human vaccine based on the Sterne strain was developed at the Microbiological Research Establishment (the predecessor of the Health Protection Agency) in the late 1950s and first made available for human use in 1963. This research center produced batches of the vaccine every year or two for protecting workers in the wool, worsted, hide, and fertilizer industries.[117]

Most of the vaccines used for humans after 1960, though based on Sterne and other avirulent strains, contained no actual bacilli. Directly following World War II, the British bacteriologist G. P. Gladstone published a paper demonstrating that the necessary component for a successful vaccine was present even in liquids from which all the cells had been removed. A group of scientists focused on one protein made by *Bacillus anthracis,* the protective antigen, in a series of experiments conducted at the

Biological Warfare Laboratories of the U.S. Chemical Corps at Fort Detrick, Maryland. George G. Wright and colleagues published a series of papers detailing their development of a vaccine that used alum to precipitate the protective antigen and separate it from anthrax cultures.[118] By the mid-1950s, British scientists also had standardized the modes of production in the laboratory for a bacillus-free vaccine that contained the protective antigen.[119] These vaccines could not cause the disease, since they lacked the bacilli. A flurry of clinical trials in animals (especially monkeys) followed, and eventually bacilli-free anthrax vaccines were offered and administered to people in at-risk populations.[120]

In 1957, American mill workers in Manchester, New Hampshire, served as the test population for the only openly documented clinical trial that supplied "hard data on the effectiveness of the vaccines in humans."[121] The Arms Mill workers processed goat hair imported from Pakistan, Iraq, Iran, and India. Like their historical counterparts in Walpole, Glasgow, and Bradford, some of the Manchester workers had periodically experienced cutaneous anthrax. Those who had not previously contracted anthrax were offered the chance to participate in a trial in which they received vaccine or placebo injections. The vaccine trial, using vaccine that contained the protective antigen but no bacilli, began in May. While it was ongoing, nine workers came down with inhalational anthrax between August and November. The scientists reported that they then offered the vaccine as a protection to all workers, effectively ending the clinical trial. Subsequently, no new cases of inhalational anthrax appeared in that mill. The 1957 vaccination trials remain controversial, with some observers contending that the trial benefited the scientists and not the workers.[122] We return to this controversy later in this book, but for present purposes, the history of anthrax in this mill highlights how the bacillus became domiciled in the factory and in workers' bodies—and how both became important sites of scientific work. Altering the internal environment through prevention and treatment of anthrax in exposed people increasingly became the preferred mode of public health control.

The possibility of completely controlling anthrax became even more tantalizing with the discovery that antibiotics could kill the bacillus in its victims' bodies. The development and mass production of streptomycin in the 1940s changed the therapeutic landscape completely for anthrax, as with so many pathogenic microorganisms. By the late 1940s, anti-anthrax serum was not even manufactured in some nations (including the United States), having been replaced within a couple of years by streptomycin.[123] One American physician, Herman Gold, remembered later how he had treated his patients early in his career with sera, then salvarsan and similar compounds, and finally with streptomycin and other antimicrobials in the 1950s. By about 1960, Gold confidently predicted that anthrax would soon no longer be a problem for humans. Even if a person became infected, antimicrobial therapy would eliminate the bacilli in the victim's body.[124] The other major component of anthrax control, creating a buffer of vaccinated and infection-free domesticated animals, would slowly reduce the prevalence of the infection in wool, hair, and hides. Anthrax seemed to be on the path to becoming a mere memory.

ANTHRAX TRANSMISSION CONTROLLED?

As an occupational health problem, anthrax affected relatively few people, yet tremendous industrial, governmental, and scientific resources were expended in understanding and controlling it. Most historians recognize that this represented the rising power of the working class and the politicians they supported.[125] National competition, particularly between Britain and Germany, also fueled a desire to remove this troublesome problem. Yarn and textile production could contribute to the drive for efficiency—the word of the day in the early twentieth century—only if its workforce was not plagued by disease.[126] Yet there were other reasons for the attention paid to anthrax. It was a foreign disease of "foul fleeces," part of a larger pattern of looking to the East as the cradle of cholera and other devastating diseases in the nineteenth century.[127] In Britain, the newly recognized legal status of industrial injury and the workmen's compensation system, which dated to 1903, also played a role in furthering the regulation of anthrax.[128]

Perhaps most importantly, anthrax inspired a pointed fear in workers, who suffered from blackened pustules and pneumonia and could be healthy one day but dead the next. Nineteenth-century attempts to understand and control anthrax preserved the image of *anthracis horribilis*—a new resident of the industrial sector of many nations. The inhalational form, also known as internal anthrax or woolsorters' disease, carried a high fatality rate that was widely publicized by workers and their political champions. As British politician Frederick Jowett put it in 1912, "any person who gets anthrax in its pulmonary form . . . is bound to die and no matter what serum is put into him or her, [inhalational anthrax] is fatal."[129] While Jowett somewhat overstated the fatality rate, he was only reflecting the broad understanding of anthrax as a particularly terrible disease with, as historian P. W. J. Bartrip has put it, "its own peculiar mythology."[130] This mythology had developed over centuries of experience with anthrax in livestock and the people associated with them, and the human experience with industrialized anthrax only reinforced the power of its fearsome reputation. Anthrax was a disease of animals—a disease that violated the organic boundary between humankind and the lower orders, a particular concern in the context of evolutionary debates in the late nineteenth and early twentieth centuries. Anthrax as an industrial disease engaged this set of historical and cultural frames, adding its own peculiar valence to arguments about the human place in the larger living world.

For the anthrax bacillus, industrialization and its domestication in new habitats meant tremendous opportunities. The processes of industrialization and globalized trade made anthrax a "cosmopolitan" organism in a biological sense. Beginning in the mid-1800s, populations of *Bacillus anthracis* had spread out into many more places, becoming established in new locations and among fresh, susceptible human and animal populations. In this way, human actions encouraged the ecological expansion of anthrax's habitat and shaped its evolutionary history. The strains of anthrax that traveled around the world in the 1800s have come to dominate the species' family tree by the twenty-first century, inhabiting all but one of the world's continents. Seldom have we been able to document so well the effects of human social processes on the development of a microorganism. Anthrax's domestication into urban areas around the world may have been bad news for the local human inhabitants, but for the bacillus (for the globe-trotting strains, at least), the nineteenth century qualified as an unmitigated biological success.

What has this meant for the control of anthrax as a public health problem? Anthrax was not only an early test case of ideas and practices based on germ theories, but also a prototype for the strategy of altering the internal environments of susceptible humans and animals. Researchers recognized the difficulties of purging the bacillus' spores from the external environment and developed immunity-altering treatments (such as sera and vaccines) between the 1890s and 1960s. To physicians of the mid-twentieth century, the development of streptomycin and other antimicrobials represented the epitome of control. Vaccination could create herd immunity, but antimicrobials would abolish the bacillus' killing ability within every victim. The control of anthrax promised to be within the grasp of the scientific and public health communities, if the preventive and therapeutic agents could be perfected and applied. Anthrax's ability to kill seemed on the verge of being abolished.

Just as effective therapeutics were being developed to prevent the havoc wrought by *B. anthracis,* however, it was on the ascent as a biological weapon. By the 1960s, industrial cases of anthrax in the United Kingdom had become rare enough that the Liverpool disinfection plant closed. Around the world, the vaccination of animals was slowly taking hold and reducing the active spore load in soil and on animal products. Anthrax declined in incidence, but it stubbornly resisted total eradication. The next chapters explore the reasons why *Bacillus anthracis* became a candidate for weaponization, just at the time when control and eradication of the organism seemed within reach.

Detailing this process uncovers some very important themes, not just about anthrax, but also about the scientific practices applied to microorganisms. In his book *The Double Face of Janus,* medical historian Owsei Temkin once wrote, "Science appears as a huge and complicated machinery that can be used for good or evil," and the twentieth-century history of anthrax exemplifies this tension. Like Janus, the ancient Italian deity who guarded the gates of heaven, bacteriology had two faces in the twentieth century. Janus' smiling face, if it could look back in time, would tell the familiar story of scientific advancements reducing the incidence of anthrax and other diseases. But Janus had a second face, on the back of his head—a visage of fury and sadness. If we could ask it about anthrax, this face would reveal that the same scientific knowledge was used to enhance the power of *Bacillus anthracis* to kill and disable human beings. The historical processes of domestication in the laboratory and factory proved crucial to enabling anthrax's use as a biological weapon—even as scientists were confidently predicting that the organism would become a mere historical curiosity.[131]

CHAPTER FOUR

Casualty Effectiveness
War and Anthrax

❋ ❋ ❋

It is a matter of textbook information that "N" [anthrax] can produce disease in man and a variety of domestic animals, but the relative susceptibility of various hosts to infection by [the] respiratory tract . . . is less known . . . a ten to twenty-fold improvement would be required to obtain the performance against military personnel that is now claimed on the basis of trials with sheep.

Special Report No. 9, Camp Detrick, July 1944

Not surprisingly, the wars of the twentieth century fostered the development of anthrax as a biological weapon. The more interesting question is how this process worked, and why scientists focused on *Bacillus anthracis* as the raw material for such a weapon. This chapter begins by explicitly connecting scientific work on anthrax in the nineteenth century (designed to eradicate or at least control the disease) with the development of twentieth-century biological weapons (BW). Anthrax was a leading candidate for weaponization in part because it was so well established in laboratories, scientists had so much experience with it, and the exigencies of wartime demanded the use of a familiar organism. Cultures of *B. anthracis* resided in state-sponsored scientific institutions, and medical and bacteriological training programs in several countries used this microbe as a standard organism for teaching. In short, *B. anthracis* was in the right place at the right time, available to scientists in a hurry to develop a microbe into a weapon.

The great European microbiological and pathological institutes—Berlin's Koch Institute for Infectious Diseases and Paris's Pasteur Institute chief among them—supported and educated scientists who worked on biological agents for weaponization. Veterinary institutions—London's Royal Veterinary College and Cornell University in New York, for example—often donated strains of anthrax derived from animals to laboratories trying to weaponize *B. anthracis*. In Japan, the majority of microbiologists, trained at state-sponsored institutions, were connected in some way to Japan's giant World War II–era Manchurian complex of weapons testing and development, including highly secretive experiments conducted on human beings. The major Russian scientific research institutes, dating from the era of the tsars, provided the skilled personnel for that nation's post-WWII BW development program.[1] In the service of the state, microbiology could fulfill many functions, such as contributing to the public health and controlling outbreaks of disease. Beginning in the early twentieth century, however, microbiology's service to the state in several nations also included shaping *Bacillus anthracis* into a biological weapon.

I excerpted the quotation that began this chapter from *Special Report No. 9*, a product of the United States' biological warfare development program during World War II. The "performance" it referred to was a euphemism for the power to kill. This chapter's title refers to "casualty effectiveness," which the American microbiologist and weapons developer Theodor Rosebury defined as the ability of a weaponized biological agent to kill or disable its victims.[2] As we have seen, *Bacillus anthracis* naturally possessed these attributes since the completion of its life cycle in the wild required killing its host. Anthrax was thus eminently qualified to become a disease that was also a biological weapon, since the ability to infect and kill hosts was a necessary component of a weaponizable microbe. Anthrax was only one of the possible diseases considered for transformation into BW—tularemia and plague

were other leading contenders. But anthrax, with its natural killing power, cultural legacy of fear, and well-domesticated bacillus,
led almost every weapons program's agenda by the mid-twentieth
century.

Even so, the process of weaponizing anthrax was complex. As
the special report excerpt suggests, spreading inhalational anthrax
(the most deadly form of anthrax, such as woolsorters' disease)
promised to be the most effective method of delivering weaponized *B. anthracis*. Unfortunately, scientists knew few details about
how inhalational infections actually took hold and developed in
victims' bodies. Making *B. anthracis* into a biological weapon required transforming it into a predictable, controllable killer of
even more potency than it had naturally. This was the goal of
scientists' artificial breeding programs for *Bacillus anthracis*—to
create even more highly infective and lethal strains. Finally, a useful biological weapon had to be controllable in order to protect
one's own soldiers and citizens from it (even while using it to kill
enemies). Control measures included vaccine development, research into the mechanics of precisely deploying the bacillus-
containing solutions, and studying the best formulations for
quickly and reliably infecting victims within a delineated geographic area. The process of transforming *B. anthracis* into an efficient biological weapon, deployable against human populations,
would prove to be a difficult one.

Even as they worked to develop weaponized anthrax on the
eve of World War II, scientists had doubts about how effective
such a weapon would be. In 1933, U.S. Army Major Leon Fox
singled the bacillus out as a possible weaponizable agent because
of its killing capacity and the fact that it was not contagious from
person to person, so it was less likely to rebound on its military
deployers. But overall, Fox felt that the technical problems of developing and deploying *B. anthracis* and other biological agents as
weapons more than outweighed the potential value.[3] Other scientists advocated its use. Usually these proponents had already conducted research with spore-forming bacilli and felt confident that
B. anthracis could be transformed into the microbial component

of a weapon. In their hands, *B. anthracis* went through a process that intensified certain natural attributes of the organism. They created new strains with increased killing capacity and hardiness, strains that could remain infective even after being dropped through the air and exploded from bombs.

There were primarily two ways to deploy anthrax as a weapon: in shrapnel bombs (warfare use only) or as a microbial cloud generated by high-altitude bombs or atomizers (usable either on the battlefield or for covert sabotage). The first method relied mostly on shrapnel wounds and environmental contamination, caused mostly noninhalational exposure, and was not covert. (The explosion of a shrapnel bomb was unmistakable.) The second method could be surreptitiously used against civilians and animals, and explicitly used for inhalational exposure. Bomblike containers could explode high in the air, or airplane-mounted atomizers could spray a solution of infective organisms, with either method creating an invisible microbial cloud that would disperse over a wide area. For the microbial cloud method to work, users had to figure out the dynamics of atomization and inhalation: which form of the bacillus (spores or vegetative) and which liquid or semisolid carrier should be used; what kinds of particles would cause disease and death more reliably when inhaled; and how the particles should be disseminated to avoid infecting the army itself. These were the problems that scientists working with anthrax concentrated on during WWII, and that spurred the development of large state-run BW programs. While rogue usage of these weapons continues to concern us, and despite widespread acceptance of the principles of the Geneva Convention after WWI, state sponsorship was the cradle and nursery of developing modern BW.[4]

The close connection between scientific institutions and covert weapons research set up a conflict of principles for the scientists involved. Weaponizing a biological agent was the moral opposite of what biomedical scientists usually did, which was to reduce the killing power of microorganisms to preserve human lives. In the private documents that survive from the WWII era, American and British scientists engaged in surprisingly little soul searching, and they did not write publicly about their qualms until after the war. Then they sought to explain how difficult their task had been. Writing in 1949, American BW researcher Theodor Rosebury called the work "upside down" bacteriology, because "BW sets out to produce disease."[5] In 1951, General William Creasy, of the Chemical Corps Research and Development Command, acknowledged scientists' difficulties when he defined biological warfare as "public health and preventive medicine in reverse."[6]

Scientists who worked on BW did so for various reasons: patriotism, fear of enemy capabilities, and even (amazingly) a feeling that death from a biological agent would be more humane than alternatives such as bullets and starvation. As the latter argument went, covert attack would happen invisibly and peacefully, and if the disease was severe enough, the victim would not suffer for long.[7] *B. anthracis* weapons could also be directed at animals, removing the moral objections of directly infecting humans (although the contaminated environment and food supply would obviously place people at risk). For most scientists, feelings of patriotism and the desperation of wartime circumstances trumped their ethical concerns. More ominously, some scientists who worked on BW viewed their enemies' civilian populations as somehow deserving of attack. This attitude was carried to an extreme in Manchuria, where many Japanese workers in the biological research and testing units referred to their Chinese, Mongolian, and Russian human victims as *marutas,* or "logs." The scientific leaders of the Manchurian units expressed few regrets during or after the war for the thousands of deaths of people they considered racially inferior and enemies. Nonetheless, some Japanese scientists have testified that they did everything they could to avoid being sent to work in Manchuria, feeling uneasy about what they had heard was going on there. Those who attempted to resign after their arrival were told that they risked execution if they refused to participate in the BW work.[8]

The ethical picture was thus complex, with parameters that varied from place to place. Although American and British workers did not have to face what the Japanese scientists did, they were still subject to practical problems. Scientific careers depended on the attribution of credit for experimental findings, and credibility was a scientist's most prized professional possession. Anyone doing mostly top-secret work was necessarily prohibited from talking to colleagues or publicizing results at conferences and in journals, and thus had largely removed him- or herself from the open scientific community and from gaining credit for research.[9] As Rosebury pointed out in 1949, a researcher's prewar work would often be taken up and moved along by other scientists—sometimes effectively ending careers or at least lines of research.[10] Scientists experienced friction with military commanders and stress due to time pressure. They exposed themselves to the deadly diseases they studied. The decision to work on biological agents was not made lightly, and the disadvantages must have been in part alleviated by the intrinsic interest of the work. Pursuing microbiological research as war emergency work and participating in interdisciplinary groups stimulated many researchers by offering unique intellectual opportunities.[11]

What about the organism, *Bacillus anthracis?* In nature, *B. anthracis* was a relatively stable organism in terms of its evolutionary history. The familiar strains of the microorganism exhibited little biological diversity, and *B. anthracis'* genetics changed only very slowly over time. The activities of scientists working in the twentieth century altered this pattern dramatically. Much as human global trade patterns had expanded *B. anthracis'* ecological niches in the previous century, its laboratory life in the mid-twentieth century greatly expanded its genetic diversity. New branches in the microbial family meant a more vibrant population able to infect new hosts in new ways. (Scientists believe that increased biological diversity leads to better long-term survival of a species, in effect winning the evolutionary sweepstakes.) Given *B. anthracis'* natural propensity to evolve slowly, new branches in the family would normally have taken tens of thousands of years to arise. The bacillus became a standardized laboratory organism with a stable domestic existence in numerous laboratories, however; and in the laboratory, scientists could manipulate *B. anthracis* within a couple of years to create exotic mutations that might fulfill human goals (such as increasing virulence and casualty effectiveness).[12]

Thus, they unwittingly expanded the genetic diversity of *B. anthracis* in a very short period. This process had two effects: first, creating an evolutionary advantage for *B. anthracis* because its biological diversity had quickly increased, and second, skewing that evolutionary development in the direction of enhanced virulence. Whether *B. anthracis* would have developed in such a way without scientific intervention can never be known. But one thing is clear: human activities have shaped its family tree in profound ways in only the past half-century or so. Those interested in creating weaponized anthrax relied on the bacillus' earlier domestication in the lab and their own specialized scientific training in bacteriological methods. As a result of scientists' efforts to increase the casualty effectiveness of *B. anthracis* during World War II, the microbe's genetic diversity greatly expanded because the laboratory nurtured strains of bacteria that may not have arisen and survived in nature. The military needs and circumstances of the twentieth century's wars not only expanded *B. anthracis'* biological diversity, but also brought it to the forefront as a biological agent for weaponization. By the 1940s, *B. anthracis* had added some very virulent and infective strains to its family tree, and anthrax had taken on a new identity as a biological weapon. To understand how this process unfolded, we must go back in time to World War I.

THE FIRST MODERN VICTIMS: ANIMALS

The year 1915 can be considered the beginning of state-sponsored, twentieth-century BW use. Most histories of chemical and biological warfare have characterized World War I as the "chemical" war. German-manufactured chlorine, phosgene, mustard, and other irritant gases destroyed the lives and health of trench-trapped soldiers in combat areas. Less well known is the use of anthrax by German agents as a biological weapon during the war. Notably, the targets of this crude anthrax weaponization program were animals, not people: the cattle destined to feed Allied soldiers, and the horses still used for cavalry, transport, and artillery purposes. In the late 1990s, microbiologist Mark Wheelis rediscovered this episode and placed it into a global framework of German antagonism toward neutral nations aiding the Allied combatants.[13] This is a fascinating story, well worth recounting here.

The German sabotage program, the first to use biological agents on a global scale in the twentieth century, used disease pathogens that were zoonotic (transmissible from animals to humans). This meant that the zoonoses—many of which were considered rare or confined to agriculture or industry—were the diseases with which modern weaponizers got their early experience. This experience has continued to influence the selection of biological agents for weaponization to the present. In 2009, anthrax, bubonic plague, brucellosis, and tularemia—all zoonoses—headed the list of agents of highest priority and greatest danger at the Centers for Disease Control and Prevention. A century ago, the selection of microbes was based partly on the availability of the agents and specialized training from microbiological and pathological laboratories. Political factors also played a role. German officers interested in the use of biological agents as weapons probably targeted animals because they were officially forbidden from using weapons that might cause undue suffering in human populations.[14] Biological agents used against animals sidestepped this concern. They could also be applied covertly, causing their worst effects days after they had been deployed, and were easily used outside battle zones. Because draft animals and food-producing animals were still essential components of warfare around 1914, germs that caused diseases in these species were outstanding agents for clandestine sabotage.

The German sabotage program targeted a menagerie of draft animals being purchased by Allied combatants from officially neutral nations and trading partners. During WWI American horses and mules, Norwegian reindeer, and animals from Argentina, Romania, Spain, and Mesopotamia all moved supplies, artillery, and even soldiers from one place to another.[15] The reindeer, for example, carried British supplies across Norway to Russia, an

Allied combatant that also depended on Romanian horses.[16] Samples of sugar cubes, intended for pack animals, carried into Norway by a German spy in January 1917 still contained viable *Bacillus anthracis* spores when analyzed eighty years later.[17] Neutral Spain and Argentina exchanged animals and supplies en route to the front, and American animals were shipped from eastern ports to France. BW were the right tools for the job of disrupting this international trade in war animals upon which the Allied effort depended.

BW worked so well not just because they targeted the living engines of the war, but also because they could be deployed covertly by individuals. The German biological program, the most well developed at the time, shows clearly that a complex web of agents, from businessmen to members of the consular and diplomatic corps, were required to keep these programs going. Many of the agents, and even the head of the German sabotage program, Captain Rudolf Nadolny, had little or no background in working with biological materials.[18] But no matter, they could depend on the trained personnel and well-established procedures developed in bacteriological laboratories. Given the long association between anthrax and Robert Koch's work, anthrax was undoubtedly being cultivated in all the major German microbiological institutes and programs at this time. Nadolny probably used cultures from German army laboratories when he sent stocks of the germs thought to cause anthrax and glanders (a debilitating horse disease that can also be deadly to humans) around the world in 1915. Some went to the German embassy in Bucharest, and others to agents hired to transport the cultures to the more remote nations. Most of these agents depended on periodic shipments of fresh germ stocks, which they inoculated into sugar cubes to be fed to the animals awaiting departure for the front.[19] The agents, stationed clandestinely in remote areas and with little or no bacteriological training, did not usually produce their own biological agents for sabotage, with one notable exception: agents sent to the United States.[20]

In 1915, Dr. Anton Dilger landed on American shores with vials of *B. anthracis* and other bacteria hidden in his belongings. Dilger would probably have looked innocuous enough. Although he had come from Germany, he had an American passport. He spoke English perfectly because he had been born in the United States after his father served as an artillery officer in the American Civil War. As an adolescent, Dilger had moved back to Germany to live with relatives. He had studied medicine in Heidelberg and Munich and completed a doctoral dissertation, for which he conducted cell culture work, at Heidelberg in 1912. By 1914, he was a German officer in the medical corps running a hospital at Karlsruhe. A year later, he was reassigned to the clandestine operations unit headed by Nadolny, and he agreed to transport anthrax cultures secretly to the United States.[21]

Once there, Dilger and his brother, Carl, rented a house in Chevy Chase, Maryland, and set up an anthrax production facility in the basement. Dilger purchased an incubator for propagating the *B. anthracis* cultures and guinea pigs for testing the bacilli's killing power.[22] The Dilgers had the training and experience to successfully grow the laboratory-domesticated strains of *B. anthracis*. Carl had worked in the brewing industry, so he was familiar with culturing organisms such as yeast. Anton had learned microbial and cell culture techniques in medical school and while completing his doctoral degree in 1911–1912. He had been trained in Germany, home of the world's most advanced system of chemical and microbiological institutes and laboratories, the place where American students with scientific aspirations sought to study.[23] Without his advanced training as a microbial scientist, Anton would have been unable to set up a successful breeding operation for pathogenic bacilli.

Although many details are missing from the historical record, it seems that Dilger made some unusual choices when cultivating the *B. anthracis*. (He died in Spain in 1918, before his clandestine activities had been discovered, and apparently he did not leave an extant first-person record of his activities.)[24] His brother and associates later said that he had arrived in the United States with live wet cultures of the bacilli in his possession; no mention was made of dried spores.[25] Thus Dilger may have brought a mixture of spores and nonspore, or vegetative, forms of the bacillus, which are much more fragile than the sporulated forms. At first, Dilger feared that the organisms he had brought from Germany had not survived, but he and his brother incubated them, and within a few months, he reported that inoculations of the organisms into guinea pigs demonstrated their virulence. He bottled up some liquid cultures—a very crude biological weapon—for hired saboteurs to use.[26]

Dilger's decision to use liquid cultures is interesting given their fragility compared with the ability of dried anthrax spores to survive transport and environmental exposure without decreasing the bacilli's infective and killing powers. Spores could have been deployed in the animals' feed or inoculated into cuts and scrapes, both demonstrated ways of contracting the disease in the early 1900s. However, it seems that methods for drying the spores were not well developed, and even twenty years later, scientists still doubted that dried formulations could retain their killing power. Dilger may have used the liquid form of anthrax cultures because the bacilli that caused glanders (another microbe he was cultivating) could be best disseminated as a liquid, and it was easier to train the saboteurs to administer all the agents in the same way. Or Dilger may have simply been applying the method with which he was most familiar, the standard procedure of putting bacterial cultures into liquid form for transport between laboratories, without thinking carefully about *B. anthracis'* ability to sporulate.

For whatever reason, Dilger passed along liquid anthrax cultures to another German agent, who then hired out-of-work stevedores to administer the formulations to animals gathered for shipment from New York, Baltimore, Norfolk, and Newport News. One of these stevedores, J. Edward Felton, later recalled that he had been given vials full of liquid, corked, to use to infect the animals: "A piece of steel in the form of a needle with a sharp point was stuck in the under side of the cork, and the steel needle extended down in the liquid where the germs were. We used rubber gloves and would put the germs in the horses by pulling out the stoppers and jabbing the horses with the sharp point of the needle that had been down among the germs."[27]

Felton and a dozen colleagues knew what they were doing— they walked along the fence of the animals' enclosure and stabbed each horse in turn, effectively injecting the liquid cultures quickly into large groups of animals, as well as contaminating the animals' food and water with the liquid. They also participated in other acts of sabotage, such as setting fires on the docks.[28] Interestingly, despite the crude and dangerous nature of this work, none of the stevedores later interviewed remembered anyone becoming ill after injecting the animals—another indication that perhaps Dilger's cultures were not very effective. ("Accidents" among personnel were one of the main ways to identify potential biological agents for weaponization.)[29] Approximately 3,000 horses, mules, and cattle were inoculated before they were sent to Allied forces in Europe in 1915 and 1916. Anton Dilger returned to Germany in January 1916, with his informal laboratory in Maryland apparently continuing to supply solutions of microbes for a while after his departure. Later, when Dilger's second attempt to get back into the United States aroused suspicion among American intelligence agents, the anthrax laboratory and sabotage program apparently ended, in late 1916.[30] Meanwhile, the American government, although officially neutral until 1917, had begun its own small-scale efforts at poison and germ warfare: experiments were being conducted with ricin, a substance made from castor beans that is highly toxic to human beings.

Most of the information about Dilger's anthrax program came out in the 1930s during the Mixed Claims Commission proceedings, based on interviews taken from surviving participants several years after the end of the war. The purpose of the Mixed Claims Commission was to assess war reparations and to investigate guilt and recommend punitive damages for acts of sabotage committed against noncombatant nations before and during World War I. The commission exhibits provided the "incontrovertible evidence," discussed later in an official report written by George W. Merck, that "as early as 1915" German agents in New York City's harbor "inoculated horses and cattle with disease-producing bacteria." (Merck, head of the American pharmaceutical company of the same name, directed the War Research Service and became chairman of the United States Biological Warfare Committee during World War II).[31] Clearly, the sabotage campaign directed at animals was part of a larger German effort that included blowing up munitions depots, setting fires, and otherwise hindering the distribution of war-related materiel.[32]

But several questions remain. First, were the anthrax attacks on animals really successful? We have little available evidence, and what we have points both ways. In the United States, veterinarians were not aware of the danger. We would expect veterinarians to have been on the front line of surveillance for such diseases, but they were not informed about the covert activities suspected of the German operatives, nor did case reports of anthrax outbreaks appear in the major veterinary journals or military veterinary communications.[33] Americans in general did fear a threat of German biological warfare, but these fears were directed at humans, not animals, and proved to be exaggerated. An outbreak of anthrax due to contaminated shaving brushes sold to civilians in 1917 was at first suspected to be an episode of German terrorism but was soon shown to be commercial in origin.[34] The German agents certainly felt that the biological sabotage programs were successful. Dilger's work earned him the Iron Cross, second class, in 1918. Another German agent, known as "Arnold," also claimed success with sickening Argentinean animals and disrupting the Spanish trade with the Allies.[35]

Besides the question of how successful these programs were, it is important to ask how the nonscientist officers leading them acquired biological agents and set up sabotage programs. Almost certainly, some German scientists, laboratories, and institutes cooperated to supply the necessary stocks of germs and trained personnel. We do not know definitively the source of the cultures supplied by Nadolny and his assistants. Indeed, due to the destruction at the end of the war of German records of covert operations, it is unlikely that we will ever know. It is possible to trace indirect connections, however, by studying the backgrounds of the scientists necessary to the operation. Besides Dilger, trained at the University of Heidelberg, three other names of scientists or physicians appeared in the few extant records: Dr. Herman Wuppermann, a.k.a "Arnold," and an Army physician, Dr. Wäle, described by a general staff telegram as having been "trained by Professor Kleine."[36]

Kleine was almost certainly Karl Friedrich Kleine, one of Robert Koch's most trusted and devoted assistants and later the director of the Robert Koch Institute for Infectious Diseases in Berlin.[37] Kleine had accompanied Koch on his investigations of African Coast fever in Rhodesia (circa 1903) and sleeping sickness (trypanosomiasis) in East Africa (1906–1907). After Kleine's return to Germany, he was a researcher and departmental head at the institute. (Koch had already resigned the directorship at this point and was spending his time traveling.)[38] The Dr. Wäle mentioned in the telegram probably spent some time (perhaps a year or two) working at the institute under Kleine. Wäle would have used bacteriological cultural and propagation techniques, and certainly would have been trained to work with replicating and controlling disease in the laboratory, a hallmark of Koch's followers at the institute.[39] This type of training would have been crucial to the scientists in the sabotage program, who were expected by their commanding officers to adapt microbial culture procedures and animal inoculation techniques to the needs of the military. In short, the instruction available at the Koch Institute and at other major German scientific institutions provided the know-how necessary to run the World War I–era BW program. Wäle, Dilger, and the other scientists were the embodied links between the great German microbiological laboratories that domesticated organisms such as anthrax and the precocious German use of biological agents against its political and military enemies. In the span of a few decades, scientists had gone from domesticating the bacillus in order to control infectious disease to developing forms of *B. anthracis* that could be used to kill animals and people.

"MICROBIAL CLOUDS": AIRCRAFT AND ANTHRAX

France also boasted a well-developed system of medical and veterinary research institutes, university laboratories, and military research facilities, along with the proud tradition of Louis Pasteur's accomplishments. Thus it is not surprising that France was another early leader in twentieth-century BW research. According to scientist Olivier Lepick, who has uncovered the little existing documentation of this program, the first Bacteriological Commission meeting (1923) was attended by Emile Roux and Albert Calmette, well-known microbiologists who served as the director and deputy director of the Pasteur Institute. The civilian-staffed Pasteur Institute and Muséum d'Histoire Naturelle joined the various military installations as active participants in the BW program. Despite Roux's early belief that it was impossible to separate offensive and defensive research, it seems that the French program focused mainly on defensive preparations against BW. The German concentration on animals and zoonotic diseases guided the French program in many ways. In particular, veterinary researchers and institutions (already closely allied to human medical ones) played key roles: the military Veterinary Research Laboratory (Paris, transferred to Maison-Alfort in 1938) hosted much of the early anthrax work; veterinarian Lieutenant Colonel Velu directed the Prophylaxis Laboratory and led the later BW research; and a pair of less well-known veterinarians, Lieutenant Colonels Soulié and Guillot, managed to hide documentary evidence of the BW research program from the Germans during the occupation. From the start, French research focused mainly on zoonotic pathogens as potential BW.[40]

French officials were aware of the German use of anthrax against animals during World War I, and they feared that German workers were covertly continuing their research after the War.[41] The French War Ministry chose Auguste A. Trillat, director of the

Naval Chemical Research Laboratory, to lead its early BW program (1921–1940). Trillat exemplified the military dependence on
scientists trained or working at the major scientific institutes. His
work also exposed the close connections between chemistry and
biology within military biological programs: Trillat was a chemist,
trained in the German chemical industry, but he had also worked
in the areas of disease control, hygiene, and public health, focusing on the disinfectant properties of formalin. By 1905, he had a
position at the Pasteur Institute in applied public health research,
so he was well connected within the French bacteriological community.[42] Trillat's training and location in the chemical research
laboratory were significant because delivery systems for BW were
initially based on the model provided by chemical weapons (later
evolving into their own system).[43]

When he began working for the War Ministry, Trillat attacked
the major problem with biological agents at that time: how best to
put them into a form easily deployable by military means. Based
on current knowledge of the formulations of chemical weapons
and his experience working with the chemistry of formalin (a liquid version of formaldehyde), Trillat focused on creating suspensions of microbes that could then be transmitted by aerosolization. Just as Dilger had done, the French researchers sought to
create agents that would survive and remain infective during field
use. In the case of anthrax, they began with suspensions (probably
liquid mixtures of the vegetative forms and some spores) and considered dry spores.[44] But the major innovation made by the French
postwar program was to advocate the use of airplanes to carry
detonating devices containing the microbes. This was a giant leap
in sophistication over the old poke-and-jab techniques used by
Dilger's stevedores. It also meant that large areas behind enemy
lines could be quickly saturated with biological agents, contaminating the water and food supplies as well as directly affecting
animals and people.

Trillat calculated that "microbial clouds," created by detonating devices dropped from airplanes, would be the most efficient
way to deploy biological agents. This delivery method likely fa-

vored the use of anthrax over other candidates because in its spore form, it was very resistant to extremes of temperature, light, and moisture. Likewise, the successful candidate organism for aerosol weaponization needed to maintain its infectivity by being resistant to adverse environmental conditions. Trillat and his associates carefully studied the influence of climate and environmental conditions on airborne microbial suspensions. Trillat believed that airborne transmission also played an important role in the influenza pandemic of 1918–1919, implying that the causative agents of influenza multiplied in the air. Historian Martha L. Hildreth considered Trillat to be a "latter day miasmatist," but his position seems to be more complex than that—a hybrid of germ and environmental ideas.[45] As Jeanne Guillemin has pointed out, Trillat saw the air over an enemy target as a sort of giant petri dish that, under the right atmospheric conditions, would effectively propagate and disperse biological agents and infect whole populations.[46]

Although other scientists worried that living organisms could not survive the heat and impact of a bomb detonation, Trillat was confident in his aerosolization scenarios. Perhaps due to the French bacteriological tradition of growing cultures in moist conditions or uncertainty about infectivity, Trillat could not envision using anything other than solutions of vegetative-form biological agents. Thus, he did not directly argue for the use of the spore form of anthrax; indeed, Trillat's influence likely biased French experts against dried forms of biological agents for quite some time. Nonetheless, *Bacillus anthracis* was one of the organisms that headed their list, and the French program of the 1920s influenced later BW programs in the selection of agents.[47] To be a useful weapon, the microbe had to be a reliable killer, and anthrax certainly possessed this quality of casualty effectiveness. In part due to its killing power and its environmental persistence, later state-sponsored BW programs would retain anthrax as one of their preferred agents (especially the dried spore form). They found that hardy spores from the laboratory could overcome the technical difficulties of aerial deployment.

But this was hardly self-evident in the 1920s, and in 1925 BW research ground almost to a halt with the ratification of the Geneva Protocol by France, Germany, the United Kingdom, and other nations, all of which would nonetheless later develop BW programs along with the nonsignatories—Japan and the United States. The French program was later reinstated, and France was the only Western power that maintained an active BW research program in the late 1930s. It did so despite being a signatory to the Geneva Protocol by justifying the program as a defense against feared German covert operations. Hitler had come to power, and the specter of airborne chemical and biological attacks validated the resumption of the French research in the minds of War Ministry officials and scientists.[48]

A retrospective report, written by the French surgeon general A. Costedoat in 1953, detailed the accomplishments of the 1920s and 1930s French BW program.[49] Much of the work concerned the ability of *Bacillus anthracis* spores, suspended in a semisolid gel, to survive an explosion and then cause infection in animals wounded by shrapnel. At the army's veterinary research laboratory, researchers tested the dispersal of anthrax spores transported by a standard aerial bomb; they also created microbial aerosols and fine mists. Especially in the case of the anthrax-contaminated shrapnel, these combinations of bacilli and bombs proved devastatingly effective at killing large numbers of experimental animals, mainly guinea pigs.[50] The shrapnel wounds, while not fatal themselves, effectively injected the anthrax bacilli into the animals' bodies, infecting and killing them. This would not be the last cooperative effort between veterinary scientists and the military in developing BW. Since most of the leading candidates for weaponization (anthrax, tularemia, and plague, for example) were zoonotic diseases, veterinary researchers were invaluable for their knowledge and access to various experimental animals.

Thus the French advanced military knowledge about the best BW delivery systems before World War II began. Research done by the French and German militaries focused on *Bacillus anthracis* as a leading candidate for weaponization for both institutional and biological reasons. This was a marriage of convenience for military delivery systems (bombs dropped from the air) and *Bacillus anthracis*. Trillat had considered other organisms along with *B. anthracis,* and experiments were conducted with bacteria (such as those causing glanders and plague) that lacked a spore form. However, *B. anthracis* best met the requirements of infectivity and casualty effectiveness, while tolerating the extreme environmental conditions of the military's favored deployment methods of explosion and aerosolization.

Despite these advances, BW researchers do not appear to have attempted to make anthrax even more versatile as a weapon. The French program transmitted the disease to guinea pigs by injuring them with shrapnel. Deep injection through a skin wound certainly worked to disseminate the disease, but it was hardly subtle or covert. Another route of infection remained to be explored, however: exposing human and animal lungs to the spores, causing inhalational or pulmonary anthrax. Here the lessons of woolsorters' disease in Britain would have been helpful, but it is unclear that the French researchers considered this historical case. When the British wool disinfection station opened in Liverpool in 1921, there was little fanfare in France (still recovering from World War I). For the British, on the other hand, the problem of occupational pulmonary anthrax remained alive in recent memory. Thus, it was in Britain, which would carry out an active BW development program during World War II, that animal tests would be conducted with pulmonary anthrax as an even more effective biological weapon.

DEATH FROM THE AIR: SPORES AND INHALATIONAL ANTHRAX

In 1888, German scientist Hans Buchner published a group of articles in the *Archiv für Hygiene* that sought to answer the question of whether anthrax spores and vegetative-form bacilli could enter the bloodstream by penetrating intact lung membranes. None of these articles indicates any interest in developing BW; Buchner was a scientist trying to understand how the bacteria worked inside the body. Buchner and Friedrich Merkel built an experimental apparatus that enabled them to aerosolize cultures of bacilli and expose rabbits and other experimental animals in a breathing chamber. They found that the vegetative-form bacilli could easily induce what they called "artificial" (*kunstliche*) anthrax pneumonia, while inhalation of spores alone caused no discernible lung pathology. Speculating that the bacilli multiplied and formed their poisons quickly because conditions in the lung were congenial to them, Buchner wrote that in a natural infection, the spores might irritate the fabric of the lungs.[51] From what we know today, it is possible that the German scientists' preparations contained spores in clumps too large to penetrate the lung passageways. Regardless of the reason, their negative results placed some doubt on the ability of *B. anthracis* in the spore form alone to cause pulmonary anthrax. This may help to explain the early twentieth-century scientists' emphasis on liquid preparations (recall that the German spy Anton Dilger used liquid solutions of anthrax bacilli to infect U.S. military animals during World War I).

When future researchers read the Buchner articles, they may have figured that the dried spores would not work as stand-alone infectious agents to be included in weapons systems. We do not know whether the French weapons researcher Auguste Trillat knew about Buchner's work, but Trillat certainly favored using solutions of bacilli for greater infectivity. Buchner's results suggested that a combination of spores and vegetative-form bacilli in liquids provided the best agents for pulmonary infection, and Trillat and his colleagues had developed exactly this type of bacillary mixture. The low-flying aircraft of the time could disseminate the bacilli by spraying them in infectious clouds to be inhaled by the victims. The hardy spores, on the other hand, could also be dropped in bombs and survive the explosion. Using both simultaneously could be one solution to the problem of balancing high infectivity and survivability of the bacilli. Before World War II, this combination seemed to have the most promise to kill and terrify human populations in the short term, while seeding the surrounding landscape for years to come.

This is important because it provides evidence that the French program of the 1920s influenced other nations' BW programs. Information about how to transform laboratory strains of anthrax bacilli into BW traveled transnationally. By the time French scientists resumed their secret investigations into anthrax-laden bombs in the late 1930s, Japanese scientists based in Manchuria had begun a far more ambitious and horrific program of testing biological agents. Since 1931, Japanese forces had occupied the Chinese province of Manchuria, renaming it the sovereign state of Manchuko in 1932. Manchuko was a Japanese puppet state, completely controlled by the commanders of the occupying Kwantung Army, until Soviet and American forces overran it at the conclusion of World War II. From 1934 until August 1945, Japanese scientists and officers tested BW such as anthrax at research stations at Ping Fan, Anda, and elsewhere. The Ping Fan research station, staffed mainly by scientists and assistants from Unit 731 of the Kwantung Army, differed from its counterpart in France in a crucial and terrible way: Unit 731 included facilities for testing potential BW on human beings. Shielded by the complete military control of Manchuko, the activities of scientists at bases such as Ping Fan remained covert until the very end of World War II and the U.S. and Soviet military takeover. In the final chaos, some scientists and many written records were lost, while others fled and escaped destruction. The Japanese Army blew up Ping Fan and other sites as it retreated, obliterating buildings but leaving behind hundreds of wandering animals (rodents, livestock, monkeys and even camels) and vast, hastily dug graves filled with Chinese and Mongolian people.[52]

In 1945, American officials, motivated primarily by the perceived need to supersede the Soviets, pursued the escaped scientists and what little evidence was left from the experiments. They also acknowledged potential benefits of Japan's experimentation to the covert American BW program, which was relatively new at the time. The Japanese investigators had recorded the effects of anthrax, bubonic plague, and many other deliberately induced diseases on the bodies of people and animals. This was a unique chance to see the results of human experiments in the natural history of certain diseases (an argument also made in the 1940s by U.S. Public Health Service personnel working on the infamous Tuskegee syphilis study in African American men).[53] The Japanese BW program was lengthy, extensive, and well funded. American officials hoped that the surviving data would provide crucial information about important problems in the preparation of biological agents, ways of disseminating agents (they tested anthrax in such vehicles as feathers and chocolates), and effective protective measures for the army using the agents. By studying the Japanese results, American researchers felt that they could jump years ahead overnight in the tedious process of testing various weapons formulations. They had been restricted to testing on animals and extrapolating to humans; now they had exclusive access to rare research on the major target species. With the Japanese data, American BW officials believed that they had found a major shortcut to their goals of understanding and controlling various biological agents.[54]

The U.S. Army gained control of the existing Unit 731 records after the war and successfully monopolized interviews of top Japanese scientists and physicians who had participated in the BW program. The Manchurian experiments were not scientifically innovative, but they were particularly morally horrific and reprehensible. Japanese scientists, led mainly by bacteriologist Lieutenant General Shiro Ishii, infected civilians and prisoners, including children and pregnant women, with anthrax and other diseases, thus torturing and killing them to identify "cheap and effective weapons." Many of the responsible scientists and their aides (including Ishii, who died peacefully of old age) escaped punishment by making a deal with the invading Americans at the end of the war: their freedom in return for providing the surviving documents and information about the research stations (including the one at Ping Fan) and the experiments conducted there. American military officers up to and including General Douglas MacArthur agreed that Ishii and the other scientists should have immunity from war-crimes prosecution, thus expunging from the historical record the suffering and deaths of thousands of people that would have come to light during a public trial.[55] As one U.S. Army memo stated, "any 'war crimes' trial would completely reveal such data to all nations . . . it is felt that publicity must be avoided in the interests of defense and national security of the U.S."[56] The decision did preserve the data for American eyes only.

Most of the extant information from the Japanese program—if it still exists—remains classified at the top-secret level and has been unavailable to historians for the past fifty years. John Powell, editor of the *China Monthly Review,* had lived in China and worked for the U.S. Office of War Information during the war. Powell doggedly pursued the story of what happened to American-controlled information about the Japanese BW program. He obtained some of the crucial evidence through the American Freedom of Information and Privacy Acts.[57] In 1981 he commented on the extreme secrecy, saying that "Japan's desire to hide its [biological weapons experiments] is understandable . . . The American government's participation in the cover-up, it is now disclosed, stemmed from Washington's desire to secure exclusive possession of Japan's expertise in using germs as lethal weapons." Powell's disclosures were explosive because they asserted that American prisoners of war were among the "human guinea pigs" used to "study the immunity of Anglo-Saxons to infectious diseases."[58] Powell's findings kept alive the memory of Japanese BW programs during the war. Besides John Powell, British reporters Peter Williams and David Wallace and American historian Sheldon H. Harris, among others, have interviewed elderly informants and delved into obscure archives in China, Japan, and the United States to piece together more of the story. History has condemned the immunity deal. As editors and contributors to the *Bulletin of the Atomic Scientists* wrote in 1980, "Any reader with a sense of justice and decency will be nauseated, not only by these atrocities, but equally so by the reaction of the U.S. Departments of War and State," which had allowed the perpetrators to go unpunished.[59]

Japanese scientists who fell into Russian hands at the end of the war were not as lucky as their colleagues granted immunity by the Americans. At Khabarovsk in December 1949, they testified about their participation in human experiments on Chinese, Russian, and western prisoners and were sentenced to years in Soviet forced labor camps. Some survived to return to Japan, but few were willing to talk about their wartime research. The Khabarovsk trial transcript was strictly guarded by Soviet officials (for reasons similar to the Americans' desire to keep any usable information secret); only judges and high-level officials were granted access. Bert V. A. Röling, one of the judges in the International Military Tribunal for the Far East after the war, later recalled how he read the Khabarovsk transcript with shock and disbelief. (This transcript, still not widely available, is the source of much of what is known about activities at Ping Fan.) Writing in 1981, Röling asserted that the American public did not know about the human experiment allegations or the deal made with the Japanese scientists captured by the U.S. Army. Röling felt that the actions of the War and State Department agents (and the American scientists ordered to assist them) represented "the danger of moral depravity, in peacetime, within the circles that have the instruments of military power in their hands."[60] To be sure, the moral calculus is quite different now than it was in the 1940s. American officials at that time felt that they were guided by political and military necessity, and they saw acquisition of the experimental results as worth the price.[61]

And what of these prisoners' lives at Ping Fan—what was the information the Americans so desperately wanted at the end of the war? Powell states, "U.S. biological warfare experts learned a lot from their Japanese counterparts," citing a Camp Detrick document that called the Japanese data "invaluable."[62] Civilians have limited access to these documents; nonetheless, it is possible to piece together some analysis from available sources. The Japanese data were voluminous and decisive on some questions. Available analyses—Sheldon Harris' chilling and authoritative historical accounting, Powell's research, Williams and Wallace's interviews—provide plenty of evidence that American BW researchers were eager to obtain any results they could from BW experiments on humans.[63] But what specifically did they obtain? The surviving

documents and narrative accounts include a great deal of information about the natural history of diseases and conditions caused
by different types of exposure to different agents, such as anthrax
and plague bacteria. For each prisoner who suffered through an
experiment, researchers carefully recorded the clinical course of
the illness and pathological data upon death. Likewise, they carefully noted the results of weapons delivery tests, such as those
conducted with anthrax shrapnel bombs blasted at people tied to
stakes nearby.[64]

In these experiments, Ishii's unit seems to have stuck with con-
ventional BW agents and technologies. Their major innovation, if
it can be so expressed, was in using humans as experimental subjects. The anthrax experiments are a salient example. Like the
French before them, the Japanese scientists conducting anthrax
tests relied largely on artillery-inflicted shrapnel wounds to infect their experimental subjects. In one experiment, ten people
were tied to stakes with their heads and torsos padded but with
legs and buttocks exposed. Anthrax-containing bombs exploded
nearby, inflicting nonfatal wounds that transmitted the bacilli. As
with the French guinea pigs, the human test subjects became infected and died of anthrax, with the course of their disease and
sufferings documented by scientific observers.[65] These tests and
others yielded crucial data unique to human victims, particularly
the dosages necessary to produce infection and deaths in 50 percent or more of exposed subjects.[66] Without these numbers, an
efficient antipersonnel weapon based on *Bacillus anthracis* could
never have been developed.

The Japanese researchers largely replicated and perfected available techniques and confirmed existing ideas. In other areas, such
as the development of delivery systems, Japanese advances were
not a great improvement over already available technologies. The
anthrax bombs, named the HA-type bomb, could be used for
high-altitude deployment from airplanes. The infected shrapnel
would cause direct infections and infect the soil with spores for
years to come.[67] Ishii also tried a type of ceramic bomb and liquid
cultures to create bacterial clouds (as Auguste Trillat had envi-

sioned before him). High-altitude paper balloons could also have effectively disseminated anthrax bacilli or spores, but when the Japanese army deployed some of these balloons over North America in 1944–1945, they sent them unarmed. Ishii's researchers tested various items, foods, and liquids that could be used for sabotage purposes; some of this information was used by the Americans later. Overall, however, the understanding of how best to weaponize anthrax did not advance greatly. Historian Kei-ichi Tsuneishi has concluded that Japanese efforts in Manchuria failed to develop an efficient anthrax delivery system by the end of the war.[68]

Like its earlier German and French counterparts, Japanese research on BW owed its existence to the state-sponsored institutions in which bacteriology had developed in the preceding decades. Indeed, Japan was probably the nation where basic research on microbes, research on disease prevention, and research on BW were most closely intertwined. Ishii provides a good example. Trained at the elite Kyoto Imperial University in medicine, he conducted postgraduate research in bacteriology, serology, preventive medicine, and pathology, and then investigated an epidemic of Japanese-B encephalitis. Already interested in bacteriological warfare, Ishii spent two years (1928–1930) touring laboratories in France, Germany, the United States, the Soviet Union, and several other nations, acquiring the technical and political information that he would use to persuade the War Ministry and begin his BW program back home.[69] Ishii was familiar with the literature: he had read the American Leon Fox's article, which was skeptical about BW, for example.[70] Thus, the leader of the Japanese BW program in Manchuria was trained in microbiology and bacteriological techniques, had extensive knowledge of European laboratory organization and procedure, and probably had some knowledge about the French and German weapons programs. The Manchurian program was also intimately connected to the other members of the Japanese scientific community. Most of Japan's microbiologists were associated with it at one time or another during the war.[71]

Why did scientists, many of them physicians, agree to participate in such inhumane work? Most of the civilians had been drafted or otherwise forced to serve the military under threat of penalty (even execution). The ranks of academic medical scientists and even hospital medical directors supplied Ishii with the workers he needed under wartime mobilization laws. Less well-trained workers served as technicians and orderlies. Many expressed feelings of fear, sadness, horror, and guilt over what they saw and did. "I was very shocked when I arrived and found out about the human experiments," recalled Dr. Sueo Akimoto. Although he resigned several times and attempted to leave, he alleged that other scientists did not mind the work: "Very few of those scientists had a sense of conscience . . . They treated the prisoners like animals." The logic went this way: since prisoners would have been executed anyway, they may as well "die an honourable death" by serving as experimental subjects and "contributing to the progress of medical science."[72] By calling the prisoners *marutas,* logs of wood, the workers could attempt to remain detached emotionally. All of this could be dressed up in the clothing of excitement and necessity. A speech given (probably by Ishii) in 1936 gives us another hint of how these scientists rationalized the horrors of their BW work: "The research upon which we are now about to embark . . . may cause us some anguish as doctors. Nevertheless, I beseech you to pursue this research based on the double medical thrill; one, a scientist to exert effort probing for the truth in natural science . . . and two, as a military person, to build a powerful military weapon against the enemy."[73]

In this light, the Japanese scientists (at least initially) did engage in a kind of ethical arithmetic about their activities, but a combination of the military's absolute power in Manchuko, racial ideology, and hyperpatriotism fueled a conflagration of horror that even Ishii realized had to remain secret. The zeal with which the retreating Japanese military tried to erase all traces of the human experiments at the end of the war tells us clearly that they realized the implications of the activities at Ping Fan, Anda, and the other bases engaged in human experimentation.

While some of the prisoners held on the Manchuko bases were

captured Westerners, Ishii continued his cordial professional relations with his Western scientific contacts. Along with information, Ishii may have also acquired bacterial or viral cultures from laboratories in Europe and the United States. Certainly he expected to be able to. When seeking viral cultures of yellow fever in 1939, Ishii sent an assistant professor from the Tokyo Medical College, Ryōichi Naitō, to the Rockefeller Institute in the United States to request both avirulent and virulent strains. The request was denied. Naitō, who had just spent a year and a half at the Koch Institute in Berlin, mentioned that upon his return to Japan he would probably be posted to the field service in Manchuria (likely Ishii's unit). His American hosts did not know about the research activities in Manchuria and did not become suspicious about repeated requests from Japanese bacteriologists for yellow fever cultures until the end of 1940.[74] This episode highlights just how ordinary it was, even during turbulent times, for scientists around the world to visit and exchange information and materials freely between laboratories. The Rockefeller Institute, which normally would have supplied material to another nation's scientists, denied the Japanese requests for yellow fever strains because League of Nations resolutions prohibited the introduction of the virus into Asian countries for any purpose.[75] But the Japanese scientists' visits and requests were otherwise viewed as normal professional courtesy and demonstrated the degree to which these practices were embedded in the European and American scientific communities.[76]

Thus it is not surprising that (aside from the human experimentation) the Japanese wartime research on BW moved along the same lines as the established work in Germany and France. Ishii viewed anthrax in the same way as other weaponizable bacteria: put it in bombs, drop them from airplanes, and depend largely on the cutaneous or intestinal route to produce anthrax infection (although it was the French who had best developed the shrapnel method of infection with anthrax by the end of 1939).[77] Ishii's unit apparently did not develop an effective formula of inhalational spores, in part due to precedent (again, we do not know if they were familiar with Buchner's work), and perhaps in part due to lack of time and imagination. It would take an innovative British group of scientists, familiar with woolsorters' disease and willing to try new ways of taking advantage of anthrax's unique properties, to figure out how to weaponize anthrax most effectively, in its inhalational form.

MAKING WEAPONIZED ANTHRAX: TRILATERAL COOPERATION AND TARGETED CASUALTY EFFECTIVENESS

The official British BW program began in 1936 with the establishment of a Defence Ministry subcommittee on bacteriological warfare charged with evaluating defensive options should Germany use BW. Intelligence reports variously credited the Germans as having developed glass capsules, bombs, and sprays of anthrax bacilli to be delivered by airplane.[78] In 1940, a group of about a dozen scientists led by bacteriologist Paul Fildes had begun covert research in a portion of the Chemical Defence Experimental Station facility at Porton Down. They focused largely on *Bacillus anthracis,* which by the 1940s had become a standard organism, domesticated to the laboratory, "well-known and understood."[79] The most well-known project completed there was the development of cattle feed (cattle cake) contaminated by anthrax that was intended to be used against German livestock, but Fildes's team also developed an antipersonnel anthrax bomb (slated for production in the United States). These modest but significant achievements meant that, as historian Brian Balmer has asserted, the British BW program was probably the first to establish that "workable" anthrax weapons could be constructed.[80] (By "workable," Balmer meant that the weapons had all of the necessary attributes: infectivity, casualty effectiveness, availability, and easy transmission.) The British program also influenced the American program, which at its inception in February 1942 did not even include *B. anthracis* on its list of likely weaponizable agents. After consulting its allies, the War Bureau of Consultants named anthrax in their second report (June 1942) as an agent that should be studied.[81]

In the wake of the Pearl Harbor bombing in December 1941, the British, Canadian, and American research programs on BW had formed a close trilateral partnership designed to create offensive and defensive knowledge of BW to retaliate and defend against Japanese and German biological attacks. This trilateral arrangement aimed to compensate for the inability of any one of these nations to maintain the research, production, and testing necessary to surpass the likely knowledge of the enemy.[82] Under this unique umbrella, biochemists, pathologists, veterinarians, and microbiologists worked together and traveled frequently to consult with their foreign colleagues. American veterinarians Harry W. Schoening (Bureau of Animal Industry, U.S. Department of Agriculture) and William Hagan (dean of the Cornell Veterinary School) were particularly important to the cooperative development of anthrax weapons against livestock. Schoening was a member of the joint U.S.-Canadian commission that administered the covert Canadian weapons facility at Grosse-Île, Quebec (an island in the St. Lawrence seaway that had previously served as a quarantine station for immigrants from Ireland and the British Isles). The director of the American BW program, George W. Merck, also sent Schoening and Hagan to Porton Down to work on the British effort to develop anthrax-laced cattle feed.[83] American veterinarians also participated in British field tests of *B. anthracis* at Gruinard Island off the coast of Scotland, a fact that came to light after the war had ended.[84]

For our purposes, the crucial accomplishment of the British and trilateral BW program was to develop the forms of anthrax bacilli that would be sprayed or delivered by aerosol bombs, and that could infect people and animals who inhaled it. Unlike the cluster-bomb-type delivery systems, which relied on creating infected wounds, the inhalational form of the bacilli could be covertly delivered as an aerosol—the microbial clouds that Auguste Trillat had envisioned twenty years before. But neither Trillat nor his immediate successors had been able to successfully infect experimental animals with sprayed bacilli. Thus, by the early 1940s, several questions remained unanswered. Could the hardy spore form of *B. anthracis* reliably cause infection when inhaled, and if so, what was the optimal particle size? How far could such particles travel? How much of a dose was required? Which strains of the bacilli were best to use? These and other questions about inhalational anthrax fascinated one of the Porton scientists, David Henderson, who devoted a large part of his career to answering them.

The work of David Henderson and others was an important turning point in the history of how anthrax became a biological weapon, because they developed the highly infective, deadly spore-based formula of *B. anthracis* designed to be inhaled by populations of people and animals. At Porton, Henderson and his colleagues created the material culture of experimentation necessary to formulating inhalational anthrax: special strains of bacilli, standardized testing procedures, an innovative test-chamber apparatus, and quantitative methods for analyzing the bacteriological data. Henderson's scientific background made him an ideal candidate for his wartime position, and it steered him toward the development of aerosolized *B. anthracis*. During the 1930s, Henderson had worked with harmless members of the *Bacillus* family (cousins of *B. anthracis*) and thus had experience with manipulating spore-forming bacteria in the laboratory. He joined the new research group at Porton in late 1940 and began working on aerosols of *B. anthracis* and other pathogenic organisms that could infect humans and animals.[85]

A New Family Tree for B. anthracis

As other anthrax investigators had done before them, the Porton group began by creating new strains of *B. anthracis* by *passaging* bacilli through the bodies of experimental animals to intensify certain characteristics. Passaging consisted of injecting bacteria into animals' bodies and then recovering and growing subsequent generations of microbes. Passaging could genetically alter the population of bacilli, presumably due to intensive selection pressure within the animals' bodies. But would the microbial population become more or less deadly? Previously, as in the case of

Pasteur's vaccine experiments, researchers had sought to weaken the bacillus' killing power, so they passaged bacterial populations through animal species other than the ones they were trying to vaccinate. (Guinea pigs served well to create vaccines for sheep and cattle, for example.) The new, weakened strain of the bacillus could be the basis of a vaccine against the disease. Laboratory groups trying to develop BW, however, such as the Porton scientists, had just the opposite goal: they sought to increase the virulence of the bacilli and to enhance infectivity. They thought that passaging through several animals of the same species could render some strains of the microbial population more deadly to that particular animal species. They used monkeys as a substitute for humans. After passing through an animal's body, the bacilli that were recovered represented something new: a new strain that behaved differently from its ancestors. A newly grown colony of bacilli became a new strain when it responded differently from its parent to biochemical, microbiological, and other tests—and when it proved more deadly to experimental animals.[86]

Lacking access to wartime laboratory records, we can extrapolate a rough outline of the processes used at Porton by putting together information from open-source work published later. Porton scientists began with strains of *B. anthracis* already domesticated to the laboratory and encouraged them to mutate by applying radiation or heat, or by passaging them through guinea pigs, rabbits, ruminants, and rhesus monkeys. The variations created on the Vollum strain of anthrax, the eventual favorite for the trilateral allies' BW research, provide a good example. Highlighting how little we really know about the sources of *B. anthracis* strains, there are two origin stories for the original Vollum bacilli. In the first, Vollum came from a cow and was collected and passed on to Porton by the Royal Veterinary College in London.[87] However, recollections of personnel and one official record cite the strain as having originated with a case of anthrax in an Oxfordshire cow, with samples sent by the local veterinarian to the diagnostic laboratory and then into the lab of R. L. Vollum, the Oxford microbiologist. From there, this strain, Vollum 14578, came to Porton Down, probably through the request of Paul Fildes, the scientific director of the BW program.[88] Regardless of which story more accurately reflects Vollum's origins, by 1942, when the U.S. program at Camp Detrick (later Fort Detrick, located in Maryland near Washington, D.C.) got under way, the Porton lab was able to donate cultures of the Vollum strain. Later, the British workers also shared a form of the Vollum laboratory strain that they had altered to make it more virulent when inhaled—M-36.[89] David Henderson and his colleagues created M-36 Vollum by passaging the original Vollum strain through monkeys. The monkeys were infected by the respiratory route, thus giving the best chance of creating a strain of *B. anthracis* that functioned as a reliably infective aerosol for inhalation.[90] The new strains had biochemical and other attributes that differentiated them from the parent strains, but not all newly mutated strains behaved the way the researchers wanted them to.[91] By testing strains on scores of experimental animals, they could select those that had the characteristics they were looking for: increased virulence and casualty effectiveness by the inhalational route. M-36 Vollum proved to be a star. So reliably virulent was M-36 and other Vollum derivatives that researchers used these strains for decades afterward to test the effectiveness of the various vaccines they developed.[92]

BW researchers also knew that they had an especially useful strain when "laboratory accidents" occurred, and allegations persist that research programs obtained some strains of bacteria and viruses from the bodies of infected laboratory workers. In the 1980s, the Soviet BW scientist Ken Alibek, who defected to the West, changed his name, and wrote a book about his experiences, recalled a nasty strain of Marburg virus named for Nikolai Ustinov, a laboratory worker who had died of it. "No one needed to debate the next step [in the research process]," wrote Alibek. "Orders went out immediately to replace the old strain with the new, which was called, in a move that the wry Ustinov might have appreciated, 'Variant U.' . . . [At] the end of 1989, Marburg Variant U had been successfully weaponized."[93] The Vollum strains sent by the British to Camp Detrick in the early 1940s also got some unexpected American upgrades. Interviews of former Detrick scientists by Baltimore journalist Scott Shane state that three Detrick workers became infected with Vollum, two by the respiratory route (Bill Boyles and Joel Willard) and one cutaneously (Bernard V. "Lefty" Kreh). Boyles and Willard died; Kreh spent a month in the hospital and survived. Bacilli isolated from all three victims were then incorporated into the research program. "Lefty's strain was rather easy to detect," researcher Bill Walter remembered. "When a colony of bacteria grew on growth medium, it came out like a little comma, perfectly spherical." The strain was named with Kreh's initials: BVK-1, or LK for short, on the scientists' log sheets. Another researcher remembered that the American-generated Vollum 1-B strain was isolated from Bill Boyles; this strain was used along with Vollum M-36 in the late 1950s at Fort Detrick.[94] Considered to be "highly virulent," Vollum 1-B joined a growing family of laboratory-produced anthrax strains in the mid-twentieth century (table 1). Laboratory accidents helped to distinguish weaponizable strains from the others, because, as Rosebury put it, "it is a safe bet that any infective agent which can be handled in laboratory animals and cultures without infecting its handlers does not have the high infectivity that BW requires."[95]

In the process, *B. anthracis* as a species changed genetically much faster and more drastically than would have occurred in nature. Because *B. anthracis* was (and is) widely believed to be very genetically stable, even the few extra strains created in laboratories around the world represented a major change in the family history of this bacillus. The profusion of laboratory-raised strains had the unintended effect of increasing the genetic diversity for *B. anthracis* in the mid-twentieth century. The ability of humans to artificially alter the composition of a species is nothing new (domesticated plants and animals are the most obvious results). However, we bear the responsibility and consequences for what we create. Modern biology associates greater genetic diversity with greater evolutionary success. *B. anthracis*' evolutionary success comes at the expense of its animal and human hosts. In providing *B. anthracis* with the opportunity to increase its genetic

Table 1. Table of selected anthrax strains in allied laboratories, 1920–1946

Strain name	Geographic origin	Year isolated	Species of origin
95	U.S. (Virginia)	1932	bovine
107	Haiti	1943	human
108	U.S. (Maryland)	1939	bovine
109	U.S. (North Carolina)	1939	bovine
116	U.S. (Iowa)	1933	canine
117	U.S. (South Dakota)	1937	bovine
205	South Africa	1942	caprine
M-36	UK	ca. 1942	derived from Vollum
Vollum	UK	ca. 1940	cow
Vollum 1-B	U.S. (Maryland)	ca. 1941	derived from Vollum
205	South Africa	1942	goat
57	South Africa	1946	goat
1928	U.S. (Iowa)	1925	cow
G-28	South Africa	1939	[no data]
BA-1007	U.S. (Iowa)	wartime (circa)	
#39	U.S. (unknown)	wartime (circa)	
#26	U.S. (Kansas)	wartime (circa)	
#37	U.S. (Michigan)	wartime (circa)	
#47	U.S. (New York)	wartime (circa)	
#20, #28	U.S. (Ohio)	wartime (circa)	
#25	U.S. (South Dakota)	wartime (circa)	
#38	U.S. (Texas)	wartime (circa)	

Sources: Sidney Auerbach and George G. Wright, "Studies on Immunity in Anthrax IV. Immunizing Activity of Protective Antigen Against Various Strains of *Bacillus anthracis*," *Journal of Immunology* 75, 2 (1955): 129–33 (table on p. 130); Stephen F. Little and Gregory B. Knudson, "Comparative Efficacy of *Bacillus anthracis* Live Spore Vaccine and Protective Antigen Vaccine against Anthrax in the Guinea Pig," *Infection and Immunity* (May 1986): 509–12 (table on page 510); Paul Keim, Abdulahi Kalif, James Schupp, Karen Hill, Steven E. Travis, Kara Richmond, Debra M. Adair, Martin Hugh-Jones, Cheryl R. Kuske, and Paul Jackson, "Molecular Evolution and Diversity in *Bacillus anthracis* as Detected by Amplified Fragment Length Polymorphism Markers," *Journal of Bacteriology* (Feb. 1997): 818–24 (table on p. 819).

diversity in the direction of greater virulence, the scientists at Porton and Detrick altered the human relationship with *B. anthracis*. They made the bacillus an even more formidable enemy to humans and our domesticated animals.

They accomplished this by stocking their laboratories in the 1930s and '40s with bacilli isolated from sick animals and people, growing the bacilli, passaging them through animal (and sometimes human) bodies, subjecting the bacilli to extreme conditions, and selecting bacterial colonies showing mutations. Although often unaware of the exact mechanism, scientists able to control virulence or make vaccines essentially altered two of the organism's components: its capsule and the toxins produced by the bacillus. The alterations were small—extremely small. Indeed, until recently scientists did not possess reliable technology to distinguish the DNA sequences of different strains.[96] For much of the past 150 years of *B. anthracis*' laboratory life, scientists instead differentiated strains based on small differences in biochemistry, behavior, resistance to antimicrobials, and the appearance of cultures. More importantly, different strains exhibited differences in killing power in experimental animals—vaccinated, unvaccinated, and of various species. This was the major criterion used by anthrax researchers during World War II, who engineered strains of the bacillus to magnify this killing power.

Testing B. anthracis' Killing Power

After choosing their preferred strains of *B. anthracis* in 1940–1941, the Porton group used their collective experience to build an experimental apparatus that would enable them to determine the most infective particle sizes and dosages. H. A. Druett and K. R. May had already created an aerosol chamber that could spray spore clusters. David Henderson and his colleague Donald Woods developed a chamber that was more easily operated, could generate single-particle aerosols, and produced clouds of sufficient concentrations. Henderson also developed the *impinger*, the part of the apparatus from which samples of the aerosol cloud were taken. These samples showed the concentrations of spores, which when multiplied by the volume of air inhaled by an animal in the chamber, yielded a total spore dose.[97] Henderson's apparatus was used by the whole group of Porton researchers interested in inhalational anthrax.[98]

Known affectionately as the *piccolo,* Henderson's apparatus exposed animals to *B. anthracis* aerosols without endangering laboratory personnel. (Henderson did admit that "an occasional organism may escape"—not highly reassuring. Lab workers wore substantial protective clothing as a precaution.)[99] The chamber apparatus also ensured homogenous particle size. Henderson's apparatus would only produce clouds of single organisms; Druett and May's apparatus was used for clusters of spores, which enabled the researchers to figure out how spore size correlated with infectivity. By studying the lungs of monkeys and other animals exposed to different aerosols in the chamber, Henderson and his colleagues sought to understand where the particles got deposited and how they gained access to the bloodstream. Although the apparatus could accommodate a variety of microorganisms in aerosol form, *B. anthracis* was "the first pathogen to choose" for testing in the chamber: its spores were hardy and it reliably inflicted definite pathology on the lungs of experimental animals.[100]

Here the connection between the Porton researchers during World War II and the British anthrax outbreaks of woolsorters' disease in the 1800s proved crucial. The Porton researchers were familiar with the British cases of woolsorters' disease from the 1800s and early 1900s; they knew that *B. anthracis* spores from contaminated wool had sickened the workers.[101] Spores worked well with modern military weapons delivery systems (such as bombs and airplane sprayers), but nonsporulated bacteria (that lacked the spores' thick walls) did a better job of reliably infecting victims. This was true of *B. anthracis.* Other microbial candidates for weaponization (such as the bacteria causing bubonic plague and brucellosis) were highly infective but could not form spores and thus often got destroyed when shot out of planes or exploding bombs. The Porton researchers had to find ways to make the hardy anthrax spores more infective and dangerous. They took a giant step toward making *B. anthracis* a usable biological weapon because they figured out how to get the spores into victims' bodies (inhalation) and how many spores were required to infect victims. The Porton scientists also knew that *B. anthracis* had been previously tested in aerosol chambers. Henderson had read Buchner's 1888 report of his experiments with atomized aqueous anthrax spores. Undaunted by Buchner's mixed results, Henderson decided to continue testing aerosolized anthrax spores.[102] With the decision to focus on *B. anthracis* spores, the British researchers veered away from the earlier work of Frenchman Auguste Trillat (who had favored solutions of nonsporulated bacilli) and greatly improved on the cruder techniques of the Japanese testing at Unit 731. They created the modern weaponized version of *B. anthracis* by conducting painstaking laboratory work with their newly designed experimental apparatus.

Henderson and Woods designed their experimental apparatus to expose monkeys, sheep, guinea pigs, and rabbits to the cloud of spores by inserting their noses into a rubber diaphragm attached to the piccolo, and the animals inhaled carefully calibrated doses of aerosolized spores.[103] These experiments must not have been easy to carry out. Even though the animals were restrained in stocks and had the masks held onto their faces, it was likely difficult to get consistent exposures and samples from uncooperative animals of various sizes. In addition, the dead space inherent in the hose-and-mask apparatus used with the larger animals also probably affected the inhalational exposures, potentially leading to wide variation in dosage infectivity.[104] Nonetheless, with the help of Porton's physicists, Henderson felt that problems of air flow, humidity, pressure, and constant concentrations of the pathogen had been reasonably well controlled in this complex apparatus (figure 4.1). Because he could control the concentration of the bacterial clouds, he could also predict an experimental animal's response to a given dose: death or recovery.[105]

It was this determination—the dose and particle size that predictably led to death—that most directly informed the development of *B. anthracis* as a weapon. Using the piccolo and chamber,

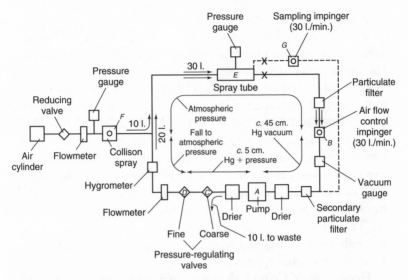

Figure 4.1. The piccolo apparatus used at Porton. David Henderson and Donald Woods' apparatus for testing Bacillus anthracis aerosols on animals. The bacilli were added to the air supply at letter F; the animal breathed the air-bacilli mixture at letter E (the piccolo); and the scientists sampled the inhaled bacilli at letter G. The rest of the apparatus regulated the air pressure and contained the airborne bacilli. This was essential, since this apparatus was an open one; to use it, personnel had to wear protective clothing and masks. David W. Henderson, "An Experimental Apparatus for the Study of Airborne Infection," *Journal of Epidemiology and Infection* 50, 1 (March 1952): 53–68, 2009 © Cambridge Journals, reproduced with permission.

the Porton group developed standardized test procedures and quantitative methods that they believed could determine dose and particle size for an infective anthrax cloud. The apparatus was a partially closed system; once the bacterial particles entered the system (letter *F* in figure 4.1), all contaminated air was supposedly contained within the apparatus or breathed in by the animal (though anyone working in the same room with this apparatus wore a respirator, gown, and gloves just in case). The animal breathed in the air-bacteria mixture at the piccolo (*E*). Researchers took samples of the inhaled air mixture through the impinger at *G* to check particle size and determine how much of the bacteria the animal had inhaled. The impinger was the crucial data-collection part of the apparatus, as Henderson said, "for sampling the cloud in such a way that accurately known amounts are collected."[106] A series of pressure valves and controllers (such as *B*) kept the pressure and flow of air in the apparatus continuous; silica gel inserts controlled humidity. After the air had passed through a filtering system (along the right and bottom sides of the diagram), the cleaned air again met the bacterial spray at *F* and headed toward the piccolo.[107]

Raw data emerged from multiple runs with the apparatus, methodically testing various animal species under many different conditions. The Porton group recorded their results on sheaves of paper worksheets. The worksheets reflected the researchers' ability to modify the formulation of each strain of bacilli (by diluting, suspending, and incubating them), the dose the animals received, and the amount of time the animals were exposed. The pressure, concentration, particle size, and droplet size of the bacterial cloud could also be controlled.[108] These raw data could support a variety of calculations, such as the optimal concentration of aerosolized spores necessary to produce a weapon. For example, the *viable spray factor* differentiated between the losses of virulent bacilli that resulted from their being damaged or killed versus their simply sticking to the apparatus while being sprayed or deployed. Calculations of the dosage for a given target animal (including humans) took the viable spray factor into account because any loss of virulent bacilli affected the bacterial concentration and volume of the bacterial cloud. The amount of infectious material needed in a weaponized cloud depended on the time of exposure, average weight and air intake of the targeted victims, and the compensation for any virulent bacilli lost on the way to the target. With their apparatus, the Porton group methodically ironed out the errors in the distribution system, identified the optimal features of the bacterial cloud, and determined the necessary dosages of bacilli.[109]

To home in on the optimal dose for infecting targets, the Porton researchers subjected experimental animals to particular conditions, then killed them and studied the result on their bodies. In their published work, Henderson's group exposed guinea pigs to a series of different concentrations of bacilli (measured in numbers of organisms per liter aerosol). After a period, they killed the animals and looked in each one's lungs for evidence of infection by the inhaled organism. For the microorganism causing brucellosis, *Brucella suis,* for example, they found that it took between 250 and 300 organisms per liter to infect 100 percent of the experimental guinea pigs.[110] Although they did not report results for *B. anthracis* in the published literature, they would have used a similar protocol of testing—and they would have used monkeys along with other species—to simulate the effects of microbial clouds on human targets. From these types of experiments, they could estimate the minimal infective doses for 50 percent of the population (the MID_{50}) and the lethal doses (LD_{50}), again using monkeys as proxies for humans.

Finally, the Porton group attacked the problem of the optimal particle size for clouds of *B. anthracis.* Knowing the optimal size of particles, and being able to control the particles in the microbial clouds, would enable researchers to create a very efficient weaponized agent. If particles were too large—clumps of multiple spores—then the clumps could not get far enough into the lungs to infect the animal. During the war, they did not know the exact mechanisms of how the bacilli gained access to the blood stream once in the lungs. Drawing on the long experience of British physicians with woolsorters' disease, they noted that "in most cases . . . a pustule has been found at the bifurcation of the bronchus" and reasoned that spores had to descend the respiratory tree at least that far (although they were careful to distinguish the "natural" woolsorters' disease infection from the laboratory simulation).[111]

Experiments designed to track the path of the bacilli in the body continued after the war. The group led by Henderson, who was by now the superintendent of the Biology Department at Porton, determined that the infectivity of the bacteria and the mortality of the test animals decreased as the particle size increased. The optimal particle size, around 5–8 micrometers, ensured that the bacilli would infect the victim, while the total number of spores deposited in one place (especially farther down into the lungs) determined death rates.[112] These results told weaponizers what they needed to know: that to maximize its casualty effectiveness, *B. anthracis* needed to be processed so that the microbial cloud contained individual spores rather than clumps. It then had to be dosed in high enough concentrations in the particle cloud to deposit as many spores deep into the lungs as possible. Even though the Porton group did not have definitive data to this effect by 1944–1945, they certainly would have collected enough information to help them decide what to put into the bomblike devices that could deliver the deadly spore clouds. They had engineered technology capable of dispensing the bacillus in precise amounts to accomplish the crucial step of calibrating the doses they would use in weapons systems.

David Henderson's personal interest in and focus on inhalational anthrax research greatly influenced work in Canada and in the United States at Camp Detrick. Henderson was a primary conduit for information between the British and North American BW development programs during the war, making several dangerous trips across the Atlantic and successfully dodging German U-boats.[113] Guided by Henderson, the North American research programs concentrated on developing weaponized *B. anthracis* using a Henderson-type experimental apparatus. After the war, American researcher Theodor Rosebury hinted at the importance of Henderson to the Detrick group: "We owe a special debt of appreciation to Dr. D. W. Henderson of the Porton Laboratories, who . . . devised the first workable equipment for exposure of animals to highly infective clouds, and who contributed invaluable basic data and many helpful suggestions during his several visits to Camp Detrick."[114] Henderson developed close relationships with his American counterparts and even married his American laboratory assistant, a WAVE officer named Emily Kelly.[115]

As the example of David Henderson's influence illustrates, Britain's BW program had begun earlier than those of its allies, and time was short; so the framework established by British re-searchers determined the parameters of the trilateral alliance's cooperative program. During World War II, the Porton group guided the British–North American BW partnership toward a re-liance on *B. anthracis* (the focus of the British program), for ex-ample. In creating and mass producing a biological weapon, each nation had its part to play. Before their alliance with the British, the Canadians had not been especially interested in anthrax, but they had expertise with other agents and with insect carriers, and they had already built strong connections with their Ameri-can counterparts. In September 1942, British program director Paul Fildes visited the Canadian group in Ottawa and revealed the results of the Porton cloud chamber experiments and field tests with anthrax spore clouds conducted that summer on Gruinard Island. The British had decided to pursue two major anthrax proj-ects: spore-impregnated cattle cakes, a type of feed for animals that could be dropped from airplanes, and 4- to 30-pound bombs designed to create a spore cloud that could kill people up to a mile away (though this was probably an overestimate).[116] In November, Fildes brought Henderson (and probably fresh *B. anthracis* cul-tures) back with him. Based on Henderson's information from the piccolo experiments, he would have explained what he knew about the characteristics of the bacilli needed to launch the most effective microbial clouds.

British BW officials aimed to use a combination of Porton's *B. anthracis* research and development, Canada's Suffield proving ground for testing weapons, and most importantly, Canadian ca-pability for large-scale production of anthrax spores. Canadian work on research and testing began shortly after Fildes and Hen-derson's 1942 visit. British information helped advance basic re-search in Canada, with at least one Canadian laboratory working to increase the killing power of *B. anthracis* strains (as the Porton group had done). The Canadians, led by the noted McGill micro-biologist Everitt George Dunne Murray, quickly geared up at a former immigration station at Grosse-Île with a goal of producing 300 pounds of anthrax spores per week—enough to load 1,500 large (30-pound) bombs.[117] During the summer of 1943, Fildes sent a British *anthrax-charging machine,* which filled Type F bombs with spore solution, to Grosse-Île. The Canadian group planned to produce the spores, charge the bombs, and test them at Suffield. However, problems with the facilities and a lack of personnel hamstrung the Grosse-Île project. Serious laboratory accidents and leaks of spores concerned Murray, while the facility was unable to attain its production goals. The Grosse-Île anthrax project produced enough of a heavy spore-suspension to charge 1,000 4-pound bombs, but the joint Canadian-British-U.S. committee guiding the spore production project decided to transfer production (and the anthrax-charging machine) to the Americans at Camp Detrick in August 1944.[118] The Canadian group continued to develop other agents, such as brucellosis, tularemia, and botulinus toxin, and a joint U.S.-Canadian team successfully created a vaccine against the cattle disease rinderpest (potentially useful to the Axis powers as a biological weapon against the Allied food supply).[119]

The U.S. BW program had begun in October 1941 with the appointment of the War Bureau of Consultants, and its wartime research and development technical director, Theodor Rosebury, followed in Henderson's footsteps and focused on creating microbial clouds. Rosebury, a microbiologist at Columbia University, had begun to worry about the possibility of enemy nations using bacteria, viruses, and other agents as BW in the late 1930s. Before the war, Rosebury had studied the oral flora of humans, which had applications in dentistry; he is considered the "father of oral microbiology" today.[120] His interest in BW arose out of his membership on the War Effort Committee of the American Association of Scientific Workers in New York. Until the United States joined the war, Rosebury had no special access to information on the subject beyond what had been published in the scientific literature. Nevertheless, in June 1942, Rosebury and his senior Columbia colleague Elvin A. Kabat submitted a remarkable report to the National Research Council in which they discussed the pros and cons of possible bacterial and viral agents for biological warfare, evaluated military delivery systems, and analyzed ways in which the nation could defend itself against a biological attack. Although Rosebury and Kabat stated that they did not advocate the use of BW, they warned, "the likelihood that bacterial warfare will be used against us will surely be increased if an enemy suspects that we are unprepared to meet it and to return blow for blow."[121]

In the 1942 report, Rosebury and Kabat had focused on airborne and insect-borne agents because of the "relative ease with which they can be disseminated and the relative impotence of sanitary measures against them."[122] Rosebury's coauthorship of this report drew the attention of War Department officials and he was appointed director of the Air-Borne Infection Unit for weapons research at Camp Detrick. Like Henderson, Rosebury was a tinkerer interested in experimental design, and he supervised construction of another version of a cloud chamber–testing apparatus. Under time pressure, Rosebury ordered a commercial germ-free chamber built by Reyniers and Son in Chicago, which was delivered to Detrick in late July 1944 and adapted to suit the microbial cloud experiments. Once a special building had been completed, tests began in earnest on biological agents in March 1945.[123] Rosebury's apparatus differed from Henderson's in that it was a more closed system, containing whole animals in boxes attached to a cylindrical chamber about three feet in diameter by four feet long. Animals moved from the exposure chamber directly into isolation rooms; workers could conduct runs without wearing respirators (figure 4.2). Rosebury quickly discovered firsthand just how difficult cloud chamber experiments were to conduct, and he gratefully accepted advice and parts for his apparatus from Henderson.[124]

The cloud chambers (there were three by the end of the war) tested various agents, including Detrick's homegrown strains of *B. anthracis.* As with their British counterparts, Detrick researchers sought to breed new strains with increased virulence. Starting with the British Vollum M-36 and N-99, obtained from Cornell University, they focused on intensifying virulence upon respiratory exposure. To maximize casualty effectiveness, the new strains

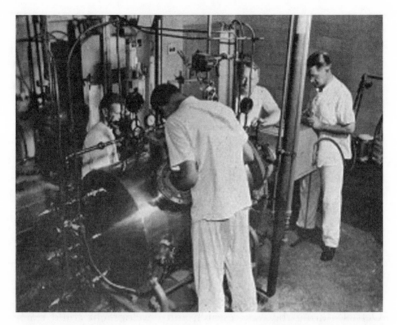

Figure 4.2. The Detrick cloud chamber. The American version of the piccolo, a cloud-chamber apparatus for testing inhaled Bacillus anthracis aerosols on animals. This was a closed apparatus, meaning that the exposed animals stayed in contained boxes, and personnel did not require protective gear (animal box is to the right in the photograph). Theodor Rosebury, *Experimental Air-Borne Infection* (Baltimore, MD: Williams and Wilkins, 1947), 35.

needed to have a predictable incubation period and cause the highest number of deaths per affected victims, or lead to the greatest duration of victims' incapacitation and convalescence if they did not die.[125] The virulence of two Detrick *B. anthracis* strains, V1 and V2, was at least twice that of Vollum M-36, and the most powerful agent they developed, *B. anthracis* strain no. 55, tested out to be three times more virulent than V1.[126] These and other deadly strains are still cultivated by laboratories around the world (with a substantial collection at the U.S. Army Medical Research Institute for Infectious Diseases [USAMRIID] Fort Detrick facility).

Creating new strains and cloud chambers for testing them required an unusual level of cooperation among scientists and physicians from various specialties—an experience that many found invigorating. Detrick scientist Murray Sanders relished the rare chance to collaborate closely with British and Canadian colleagues.[127] While directing the Air-Borne Infection Unit at Camp Detrick, Theodor Rosebury felt that the interdisciplinary approach to solving scientific problems was the most successful he had ever seen. "Cooperative research was a product of war," Rosebury wrote later, and it put "the departmental fences of many peacetime laboratories to shame."[128] The shame, to Rosebury's mind, was that many of the "great unconquered illnesses of man" and other large-scale social problems could be solved only by the application of cooperative research (or operational research, in British parlance). When "bacteriologists, physiologist, pathologists, chemists, physicians, veterinarians, botanists, physicists, engineers, and machinists" worked together, as they did at Camp Detrick during the war, research and development could reach "magnificent" attainments and goals.[129]

With such promising strains and the means to create controlled clouds of them at Detrick, the Americans hoped to begin stockpiling usable weapons in 1945. No one knew how long the war would last, and BW might be needed in the intensifying Pacific theater of the war. Weapons were tested in the United States as well as in Canada (where the Suffield testing site remained active). A remote area within the Dugway Proving Ground in Utah, Granite Peak installation, met the criteria for field testing of *Bacillus*-charged weapons. However, the mass production of *B. anthracis* and its harmless cousin, *B. globigii* (used as a simulant in weapons tests), proved to be a big problem, as it had been in Canada. A former ordnance (weapons) plant in Vigo, Indiana, was hastily remade into a BW production facility. Problems beset Vigo, however, and in the end very little weaponized *B. anthracis* was produced there. The solutions of spores seemed to lose their virulence when transferred from the laboratory to the manufacturing plant; the bomb-charging apparatus needed to be redesigned; and contamination problems continued. Construction delays meant that the Vigo plant could not be fully operational until November 1945.

The final hurdle to an American-produced *B. anthracis* weapon was political. Secretary of War Henry Stimson never approved stockpiling *B. anthracis*–based weapons, instead opting to concentrate on nuclear weapons.[130] The failure of North American efforts to mass produce *B. anthracis* meant that the trilateral collaboration never did develop weapons containing aerosolized bacilli that could be used by military commanders. By the end of the war, the British held the only usable stockpiles of an anthrax weapon: the *Bacillus*-laced cattle cakes, which were never deployed. Thus, in both World Wars, livestock would be the primary victims of an anthrax attack, and the weapons available for use were decidedly crude. Nonetheless, the feverish scientific work during World War II had taken several important steps toward making *B. anthracis* into a biological weapon.

CASUALTY EFFECTIVENESS AND "TOTAL WAR"

During World War I and World War II, scientists put *B. anthracis* through an intensive program that transformed it from an agricultural pathogen and laboratory denizen to a viable biological weapon (on a small scale, at least). Despite a continuing climate of secrecy surrounding these events, we can make out the major details of the weaponization of *B. anthracis* in the first half of the twentieth century. This weaponization was made possible by several factors: scientists already knowing how to handle it in the laboratory; the switch from developing microbial bombs to investigating the inhalational form of anthrax; and the ways in which scientists figured out how to enhance the bacillus' natural attributes and begin to marry them to military technologies.

First, had *B. anthracis* not already been well ensconced in the laboratory, anthrax would not have been selected for development by so many state-sponsored BW programs. Weaponization was completely dependent on the prior domestication of the bacillus to the laboratory: the standard procedures for cultivating it, the existence of standardized strains, and the ease with which strains could be swapped between laboratories. The microbiological institutes and the discoveries of the nineteenth century were prerequisites to creating a weaponized form of *B. anthracis*. During WWI, the German sabotage program got its strains of *B. anthracis* from the great bacteriological institutes founded there in the nineteenth century. The proud legacy of Robert Koch, who first elucidated the bacillus' life cycle and the practices necessary to cultivate it, meant that *B. anthracis* was a common inhabitant of German laboratories. Anton Dilger and other German BW agents had been trained at these major universities and institutes. They first sought an agent for use against animals, since the German high command prohibited the use of BW agents against people (although, of course, it had no such scruples about chemical warfare), and anthrax fit the bill.[131]

During WWII, the nineteenth-century industrial experience of woolsorters' disease encouraged British researchers to attempt an inhalational form of weaponized anthrax. Given more time and resources, they certainly would have succeeded. Beginning with the French researcher Auguste Trillat in the 1920s, the new paradigm of infective clouds of germs slowly replaced that of bombs.[132] However, aerosolized weapons for inhalational exposure only became effective and feasible with the work of British researcher David Henderson and his colleagues at Porton Down in the 1940s. Familiar with the grim history of woolsorters' disease, they developed the first set of experimental procedures and an apparatus that would reliably cause deadly inhalational anthrax. The key to the British innovations lay with their focus on the spore form of the bacillus, a focus explicitly encouraged by their knowledge that woolsorters' disease was spread by spores. The factory had been a site of experience with the disease, and the laboratory now became a site of production of an inhalable spore form of the bacillus. Aerosolized spores of extremely virulent strains of *B. anthracis* for inhalation became the formula of choice for weaponized anthrax and remain so today.

Delivery systems for BW were initially based on the model

provided by chemical weapons (beginning with the work of Auguste Trillat in the 1920s and 1930s). Anthrax is a particularly good example of this. During WWI and through the early years of each BW program (French, German, Japanese, American, and British), scientists focused on the idea of dispersing BW from bombs and understood BW as "poisons."[133] Anthrax got a head start as a good candidate for a biological weapon because its tough spores could withstand explosion and remain infective; it fit the paradigm built around chemical warfare. Some researchers saw BW as improvements on chemical weapons, being more covert, having a delayed effect, and being highly infective and virulent, and anthrax excelled in these areas. Anthrax took up where poisonous gas left off: its incubation period made it seem to appear out of nowhere, and it was persistent, infective, and proliferative. Like chemicals, it could be easily disseminated by airplane. BW had another similarity with chemical weapons: they were abhorred with particular vehemence (most famously by Adolf Hitler) after the horrors of gas poisoning during World War I.

Many of the scientists trying to develop BW felt the moral repugnance of their work keenly. The development of a biological weapon required the work of experienced scientists, but the training of laboratory bacteriologists and physicians focused on reducing disease and controlling the virulence of microorganisms. Biological warfare, in contrast, required scientists to strive toward different goals: *causing* disease and seeking to incapacitate as much of the target population as possible. This goal was hardly congruent with the Hippocratic Oath and the cultures of the healing professions. According to many scientists who worked on BW development during the war, only the desperate circumstances of the hostilities and the fear that the enemy would use such a weapon justified their participation.[134] Scientists such as Theodor Rosebury, who became a peace activist and member of the Pugwash Conferences on Science and World Affairs after the war, adapted only reluctantly to wartime circumstances. They were not blind to the moral implications of their actions but saw their work as necessary for the nation's defense.

These moral struggles did not prevent some of the scientists from continuing to work on BW after the war, however. These scientists' wartime efforts gave the weaponization of biological agents a new lease on life and even provided new justifications for continuing research. As David Henderson stated at the end of the war, "the overall achievement in five years of experimental study has been to raise biological warfare from the status of the improbable, where the difficulties involved were believed by many (without experimental evidence) to be insurmountable, to the level of a subject demanding close and continuous study."[135] Some scientists, such as Henderson, continued with BW research in the postwar period, justifying their work as a method of cold war deterrence that simultaneously served the needs of basic science.

Ultimately, work on *B. anthracis* as a biological weapon during WWI and WWII both enhanced its natural attributes and married them to military technologies. Scientists eager to collect, cultivate, and breed ever-more-deadly strains increased *B. anthracis'* genetic diversity. *B. anthracis* has long been viewed as one of the most genetically boring organisms we know; its genetic diversity is so subtle that only recently have scientists gained the ability to measure it. Yet differences in strains' pathogenicity are very real, and every dispersal of the weaponized strains increased the bacillus' available habitat and the ecological spread around the world of its most virulent forms.[136] *B. anthracis* appears to have been a natural killer from its ancient origins. In 2003 a summary of scientists' recent work on genetic markers showed that families of *Bacillus* microorganisms (including *B. anthracis*) required complex molecules (proteins) from insect or animal bodies, suggesting that their common ancestor was not simply a benign soil dweller but some kind of parasite that ignored plants and actively preyed on insects and animals. As Julian Parkhill and Colin Berry commented in *Nature*, "All of this serves to underline the point that, whatever crude attempts are made by human beings, the true biowarfare experts are the bacteria themselves—they are constantly ready and exquisitely able to adapt to, and exploit, any environmental or pathogenic niche that presents itself."[137] Scientists, by exponentially increasing the available niches, both engineered a weapon for themselves and helped the bacillus to expand its territory across the globe.[138]

These aspects of the relationship between researchers and the bacillus combined to ensure *B. anthracis* a place in the BW programs of the cold war era. BW were held up together with atomic weapons during the cold war as necessary components of the *total war* philosophy, a scorched-earth policy against civilians, animals, and crops that justified certain uses of biological warfare.[139] In some ways, *B. anthracis* was an ideal weapon because it destroyed life and not property—property often had to be replaced or rebuilt by the victor in a war, an undesirable military consequence.[140] However, scientists forecasted risks (beyond the moral issues) to using *B. anthracis* as a weapon. Any deployment carried the possibility that the weapon's user could be infected, and an anthrax weapon would seed the bacillus into new territories and contaminate them for the foreseeable future. In addition, as each new strain was developed, the amplification of research on virulence and vaccines seemed to be even more urgent, creating a costly imperative of weapons and defense development (a sort of bacteriological arms race). To work with *B. anthracis* was to dance with the devil. The bacillus not only was virulent and dangerous, it was also difficult to control—a theme that would appear again and again as the twentieth century yielded to the twenty-first.

CHAPTER FIVE

Resistance

Anthrax, the Modern Laboratory, and the Environment

❋ ❋ ❋

Biological weapons have massive, unpredictable and potentially uncontrollable consequences . . . Mankind already carries in its own hands too many of the seeds of its own destruction. . . . [Biological weapons are] repugnant to the conscience of mankind.

U.S. President Richard M. Nixon, 1969

In June 1942, British biological weapons (BW) researchers David Henderson and Donald Woods took a ten-minute boat ride from the northern Scottish mainland to a ruggedly beautiful local picnic spot, Gruinard Island. To field test the lethal strains of *Bacillus anthracis* developed in their Porton laboratories, they planned to dust captive sheep, the atmospheric stone ruins of crofters' cottages, and expanses of bracken and heather with various types of spore preparations. Tension hung over the scientists as they scouted the location for the tests. "We were breaking new scientific ground," as one of Henderson and Woods's group later remembered. "No-one in the western world had tried this before. So no-one knew precisely what to expect."[1]

Henderson and Woods directed a small army of soldiers, veterinarians (some American), artillerymen, and military engineers who blew up a bomb containing *B. anthracis*–laced slurry near the ground, exposing and eventually killing the sheep.[2] High-level military officials observed the events. For this and a dozen subsequent trials (including a bomb dropped from an airplane), air-sampling devices helped the scientists determine how well the spores dispersed and what the likely killing dose would be. Soldiers in protective suits burned the sheep carcasses and dynamited a cliff over the gravesite. (In the process, they were also testing the protective masks they wore. If any of them became ill, I have not seen documentation of it.) With its center heavily contaminated with resilient anthrax spores, Gruinard Island was abandoned and quarantined for another half-century—a new anthrax district called the "Isle of Death" by the locals on the mainland. Sailors occasionally spotted two ragged sheep, escaped survivors of the tests, grazing on the uncontaminated north end of the island in the dusky evenings, an eerie image of a feared place.[3]

The tests at Gruinard Island illustrated a major change for anthrax in state-sponsored weapons programs during the cold war: viewing *B. anthracis* as a strategic covert agent for sabotage and large-area coverage directed at civilians (rather than as a bomb agent on the battlefield). *B. anthracis* made a good covert agent because it took a few days for victims to become ill, making it more difficult to pinpoint the source of the weapon. Using *B. anthracis* in this way required that it be deployable as an inhalational agent, to make the deployment less obtrusive and to maximize casualty effectiveness.[4] Mechanically processing the spores for maximum inhalational infectiveness was the penultimate step in making *B. anthracis* into a biological weapon. Interestingly, the focus on inhalational forms of the agent echoed the post-WWI-era vision of Auguste Trillat, the French scientist who had envisioned microbial clouds of organisms dispersed by airplanes. By the 1950s, British and American (and probably Soviet) open-air tests with the inhalable pathogens had underscored the difficulties of using tularemia, viruses, and brucellosis in open-air tests.[5] *B. anthracis* had advantages over these other agents since it survived the exposure to sun, wind, and rain. Some strains that had lived in the laboratory for years remained infectious and hardy in the field.

The trials at Gruinard Island established the products of Porton's laboratory as the "western world's first feasible biological weapon" and demonstrated the effectiveness of the aerosolized form of *B. anthracis* (which remains the standard weaponized form today).[6] However, Gruinard Island was not the only place where BW scientists and their military collaborators created new anthrax districts at field-testing sites. Large sections of Manchuria (from the Japanese testing during the war), Dugway Proving Ground (USA), the Suffield Experimental Station (Canada), Vozrozhdeniye Island in the Aral Sea (Uzbekistan and Kazakhstan), and other areas have been environmentally altered by their use as test and disposal sites for anthrax bacilli. Returning *B. anthracis* to the field was an important step in changing its identity into a biological weapon, for practical and ideological reasons. The practical reasons are obvious: scientists needed to know how spores dispersed, whether they retained their infectiousness and casualty effectiveness, and which methods of spore preparation and weapons delivery maximized the kill rate. But the ideological reasons were equally compelling. To paraphrase historian and philosopher Bruno Latour, the bacillus would be a biological weapon only when it had done everything that the potential users of the agent (military officials) knew a weapon should do. The field trial sites became "theaters of proof"—proof that the scientists' cultures of *B. anthracis* were no longer just the cause of a livestock disease, or silent inhabitants of the laboratory, but had now become biological weapons.[7]

In the most important aspect of this identity change, *B. anthracis* was a biological weapon when it demonstrated its ability to kill, efficiently and as an *inhalational* agent, in the field. This followed from the idea that laboratory-raised and engineered bacilli were *artificial* agents that may have looked promising in the supportive environment of the laboratory but that could fail in the field. Perhaps the laboratory-raised anthrax bacilli had lost some of their "natural" components necessary for survival in the environment in the process. Under the umbrella of defensive BW investigations, scientists have continued to collect, nurture, and manipulate new virulent strains of the bacillus from the 1940s through the early 2000s.

Over the years, most nations hosting state-sponsored programs have been signatories to BW conventions and treaties banning the offensive development of biological weapons. Thus preparing weaponized forms of *B. anthracis* had to be carried on clandestinely by state-run programs in violation of international agreements, or on a much smaller scale, under the guise of defensive measures (such as vaccine development). The major state-sponsored weapons programs (USA, UK, and USSR) used both approaches for varying periods. Then U.S. president Richard M. Nixon changed the global political landscape for BW when, in 1969, he publicly announced a new American policy: the United States condemned the use of biological weapons and would restrict research to defensive purposes only. Stockpiles would be destroyed, and all production and development of BW would cease.

Nixon articulated some of the reasons for this policy change. Biological weapons were not only "repugnant," they were "uncontrollable."[8] But the Nixon administration had other good reasons for making the announcement. With the use of chemical agents in Vietnam, public hostility toward chemical and biological weapons had exploded. Nixon's advisers figured that biological weapons were redundant in an age of nuclear proliferation. Biological weapons were more easily abandoned if a nation owned a nuclear stockpile, as the United States did. In 1972, the major BW-producing nations and many others signed the Biological Weapons Convention (BWC), the most important international agreement on BW since the 1925 Geneva Convention (which the United States finally ratified in 1975). The BWC banned research, development, possession, and deployment of biological weapons—the major goals of offensive weapons programs.

Unfortunately, even the political breakthroughs of the 1970s did not signal the end of *B. anthracis'* development as a biological weapon. In 1979, around one hundred people died and many people and animals became ill with anthrax in the area around the small city of Sverdlovsk (Yekaterinburg), in what is now south-eastern Ukraine. The Soviet government claimed for years that the incident arose from infected meat in the food supply but later admitted to a leak from a military laboratory engaged in weapons preparation.[9] Other outbreaks have looked equally suspicious. Anthrax has historically been a rare disease in what once was co-lonial Rhodesia (now Zambia and Zimbabwe). After a reported average of about 11 human cases of anthrax per year, Zimbabwe suffered a major outbreak in 1979–1980, with an average of 5,375 human cases per year (and 182 fatalities).[10] This epidemic occurred in conjunction with a civil war and affected people from particu-lar racial and socioeconomic groups. Observers have long been suspicious that sabotage with *B. anthracis,* rather than natural in-fection exacerbated by the wartime conditions, caused this out-break. If so, then this was an example of what could happen when smaller military groups outside the control of the BWC and other deterrent policies acquired *B. anthracis.*

Jeanne Guillemin, sociologist and member of the team that investigated the Sverdlovsk outbreak, has summarized the prob-lem well: "Like a bad genie, anthrax refuses to be contained" by international agreements.[11] Particularly since the dissolution of the USSR and the subsequent lack of state control over former an-thrax laboratories and field-testing sites, global health officials and biological weapons experts have become increasingly concerned over the specter of virulent *B. anthracis* strains in the hands of rogue states or terrorist groups. *B. anthracis,* in certain forms, qual-ifies as a weapon of mass destruction. It is easier to obtain and use and far less expensive to develop than nuclear weapons. As one observer put it, weaponized *B. anthracis* and other biological agents are "the poor man's atomic bomb."[12]

After World War II ended, research on *B. anthracis* continued. In the United Kingdom, United States, Canada, and the Soviet Union, offensive BW programs continued for at least a few years after the end of the war. The British program refocused to be solely defensive in the 1950s, while France began another short-lived offensive program at the same time. The close U.S.-Canadian war-time relationship persisted, but Canada also slowly withdrew from offensive BW work. That left the United States and the USSR, not surprisingly, as the world's two major state-sponsored offensive BW programs between 1945 and 1969 (the year President Nixon announced the end of the U.S. program).[13] The shifting focus of these programs over time presented their employee researchers with some choices. At the conclusion of the war, scientists either left the programs, continued working on anti-anthrax vaccines or basic research using *B. anthracis,* or in some cases remained military scientists working on offensive weapons. Similar to Manhattan Project scientists after the war, many BW researchers struggled with their personal feelings about the morality of performing scientific work in the service of producing potential weapons.

From the bacillus' perspective, the second half of the twentieth century brought another tremendous increase in its ecological range (much as global trade had done in the nineteenth century). The purposeful development of new strains of the bacillus, which expanded its biological diversity, meant an ecological windfall for the bacillus when these strains escaped or were escorted out of the laboratory. *B. anthracis* is one of the most resilient environmental bacilli known to humankind. In its spore form, a strain of bacillus can persist for decades (or, some argue, even centuries) in the soil of a place where it has been deposited. Ever the opportunist, even the transmuted laboratory-generated *B. anthracis* spores deposited in these places retained the key to propagating themselves: the ability to infect and kill new hosts. As new strains got dispersed in areas around the world where they hadn't been before—isolated islands and whole regions alike—*B. anthracis* increased its ecological habitat and seeded new environments in which its life cycle could be completed. Once *B. anthracis* had been distributed in an area, it proved very resistant to removal. *B. anthracis* increasingly became an artifact rather than a natural agent in the twentieth century, but no matter how thoroughly manipulated in the laboratory, *B. anthracis* still had the power to remake landscapes into anthrax districts.

ARTIFICIAL ANTHRAX, ARTIFICIAL SCIENCE
The Secrets of Killing

The creation of biological weapons depended on the active participation of scientifically trained people and the availability of laboratory equipment. But even well-equipped BW scientists faced some essential problems with this kind of work: it was morally questionable, and it required changes in customary scientific thinking and practice. Theodor Rosebury, the director of the Air-Borne Infection Unit of the American biological warfare program during the war, later published *Peace or Pestilence,* a critical assessment of biological weapons development written for popular audiences. Rosebury articulated the difficulties for scientists in this way: "If you want to understand BW you must figuratively stand on your head."[14] Scientists were trained to approach disease-causing microorganisms by working to prevent infections and disease outbreaks. Standard scientific protocol also relied on using experimental processes to mimic the natural course of important diseases. By learning as much as possible about a microorganism's life cycle and mode of killing, for example, its interrogator could learn its weaknesses and the best ways to combat it. When trying to create a biological weapon, however, scientists needed to reverse these assumptions. A microorganism functioning as a biological weapon was *supposed* to cause disease and death, and scientists' manipulations were all geared toward these goals. General William Creasy's 1951 definition of biological warfare as "public health and preventive medicine in reverse" persisted in the scientific community critical of BW development during the cold war.[15] Martin Kaplan, the World Health Organization veterinary epidemiologist and influential secretary general of the Pugwash Conferences on Science and World Affairs, opposed biological weapons development and testing as "public health in reverse" for his entire professional life. Creasy's phrase still appears in publications about biological weapons today.[16]

In this way and others, BW research was "artificial science" that created artifacts. The artifacts were the products of the laboratory: new strains of the bacillus, vaccines, cultures, and the like. Historians of science have given us some excellent examples of the creation of these artifacts: hybrid organisms that retained "natural" characteristics even while being profoundly altered by their scientific keepers (*Drosophila* fruit flies, arguably the most studied organism in the history of genetics, come to mind).[17] Far from losing their natural characteristics, these artifacts represent a sort of "second nature," something that had started as a natural object that people then modified. The bacilli produced in the laboratory were things of nature containing overlying human-created characteristics.[18] *Bacillus anthracis* of the "first nature" existed in the soils of anthrax districts and in the bodies and bodily tissues of infected animals. Once those bacilli entered the laboratory and were grown in aqueous humor, on sheep's blood agar, or in other media, they changed materially (and representationally) from what they had been; human technology rendered them "second nature." Cultures that might not have survived or persisted in the natural environment did so in the laboratory—if their human keepers wished them to.[19] Cultures could also be discarded, and bacilli lacking the desired characteristics were disposed of, thus removing them from the gene pool. *Bacillus anthracis'* biological diversity had been shaped by scientists since the 1870s, and the pace of this activity only accelerated in the mid-twentieth century.

To many of the scientists engaged in working with *B. anthracis,* the processes required to create second nature bacilli were themselves artificial manipulations. For Theodor Rosebury, altering these microorganisms for use as weapons was not "normal or natural, but abnormal and artificial."[20] For BW researchers, trained in microbiology or bacteriology, working in offensive programs meant that they had to goad their organisms into being more efficient and effective killers. They got to think "right-side up" part of the time because they also sought to develop protective vaccines and other agents against these strains. Nonetheless, they realized that there were "portentous moral issues involved."[21] BW researchers daily watched the suffering and deaths of experimental animals subjected to these agents and worried about their own exposure. They knew only too well what the consequences could be if the fruits of their labors were used as weapons. The daily grind of "upside down thinking" and of "standing on one's head"—enlisting familiar practices in the service of destructive goals—created profound unease for many of the BW scientists working during the war and drove many of them to abandon working with *B. anthracis* once the war emergency was over.

Rosebury argued that once investigators were able to think "upside down," they often found that the process of researching particular microorganisms for weapons use was more straightforward than the usual research methods. In 1949, he commented that biological weapons research could "find ways to go forward where natural science falters" when the laboratory was the site of research.[22] One could choose microorganisms likely to meet the criteria necessary to become a biological weapon. If one microorganism did not work well, the researcher could simply move on to the next one.[23] In military work, microbiologists did not have to slavishly try to re-create natural conditions in the laboratory to understand every aspect of microorganisms and the diseases they caused. Their goals did not include the intricacies of the organism's ecology or evolution, physiology or biochemistry (unless it related to virulence). Instead, they created optimal conditions for growth and encouraged genetic and developmental differentiation. Then they selected strains for high infectivity, resistance, and casualty effectiveness. As long as a particular microorganism worked as a biological weapon, wrangling with other problems or questions was unnecessary. The laboratory bacillus, living under and altered by artificial conditions, represented the control made possible by these conditions.

Control of the bacillus did not translate into control of the disease, however. Scientists often had trouble duplicating natural disease under laboratory conditions. This represented a serious problem for those studying disease in the laboratory for public health reasons, in order to decrease the incidence and impact of disease. BW researchers were not constrained by these issues, however. For them, the translation between bacillus and disease, first in the laboratory and then in the field, was far less problematic. "Artificial" science benefited from an unusual ability to take detours around experimental obstructions. For example, BW researchers at first had difficulty with infecting experimental animals through the inhalational route. *B. anthracis'* life cycle set up the organism to be eaten by animals as they grazed in pastures; they could also be infected by spores settling into superficial wounds and germinating there. In this natural setting, animals seldom acquired the disease by inhaling the spores (the same was true of people exposed to it because they worked with livestock or meat). In the laboratory, however, BW researchers were not concerned with re-creating the disease by the natural means of transmission. Because inhalation served the needs of aerial dissemination, researchers focused on altering the spores of the bacillus to make them more infective by the inhalational route. Nature, they confidently asserted, could and should be bypassed. As Rosebury put it, the "malignant purpose" of biological weapons agents could "be *more* aptly served if its weapons elicited effects very different from natural disease, the better to aggravate problems of defense and to terrorize its victims."[24]

Anthrax during the Cold War: The Dual-Use Agent

At the end of the war, the United States, Canada, and the United Kingdom had to decide what to do with their BW programs. They were surprised to learn that Germany had not advanced much in this area (all three nations had used the German threat as a justification for BW programs in the first place). The new threat was the USSR, the only nation that had reportedly managed to create a human anti-anthrax vaccine during the war.[25] Creating a vaccine meant that the Soviets had an unmistakable advantage over their wartime allies. If one's own armies and civilian population could be vaccinated, then a major obstacle to using anthrax as a weapon on remote battlefields and behind enemy lines had been removed. The agent was not likely to rebound against its deployers in any significant way (unless it was used on one's home territory).

For this reason, development of protective vaccines and treatments against biological weapons remained the rationale behind defensive BW research programs. The British program based at Porton Down, for example, shifted gradually to vaccine production in the 1950s and 1960s. In the first few years after the end of the war, however, Britain continued to pursue the offensive aspects of its program (explicitly geared toward weapons development).[26] Offensive versus defensive is an important distinction, but it also masks the fact that many of the resources and practices required for offensive and defensive anthrax programs were quite similar. In general, scientists used the same materials and equipment to manufacture both weapons and vaccines, for example.[27] *B. anthracis* itself functioned as a *dual-use* agent, one that caused an infectious disease in humans and animals but that also could be developed into a biological weapon.

The first stages of working with a dual-use agent were the same regardless of the final product: scientists had to collect, cultivate, and begin to manipulate it. Only then could an agent for weapons use be produced in large quantities and tested for military deployment (steps that defensive preparations would not undergo). An agent for defensive use (a vaccine, for example) underwent several steps of attenuation (weakening, or "taming"); then scientists used virulent strains of bacilli (suitable as weapons) to test the effectiveness of the vaccine preparations. Then vaccines, too, had to be produced in large quantities and tested. Thus, weaponizable agents were (and still are) present even in defensive-only laboratories. Standard practices for raising and manipulating bacilli would continue to be used by researchers even after the abandonment of offensive aspects of the BW programs. As Piers Millet has described, dual-use agents have been particularly prominent in anti-animal bioweapons programs: "Much of the technology (both tangible and intangible) required . . . has both weapons-related and civilian applications."[28] This is not to say that scientists dissembled, hiding offensive work behind vaccine research or other peaceful investigations. Rather, the boundary between the two sets of practices has long been more fluid than it would seem.[29]

Although the wartime effort had produced no viable weapons that would disperse highly infective aerosolized spores, it had proved that this could be done. Researchers at Porton Down had accomplished this by devoting five years to "experimental study" of the organism, enhancing its infectivity and casualty effectiveness and equipping it to survive military deployment.[30] David Henderson, for one, was loathe to discontinue the research at the end of the war just when it was becoming productive. He urged the Defence Department and Chiefs of Staff to continue the program, which they agreed to do by mid-1946.

The postwar British program, reorganized and renamed, initially recognized that scientists would continue wartime practices geared toward further development of weapons. The Porton group's new name was the Microbiological Research Department (MRD). When Paul Fildes left to return to civilian life, David Henderson enthusiastically accepted appointment as the superintendent of the MRD. Henderson outlined a vigorous program that expanded the facilities and personnel of the department. He apparently had little trouble filling his available positions despite a labor shortage at the time.[31] It was clear that the MRD would continue working on weapons development as well as preventive measures and basic research. The two were viewed as inseparable. As the chairman of the Chiefs of Staff Biological Warfare Subcommittee wrote in 1950, "for the successful development of defensive measures it was essential to study the offensive field."[32]

This calls into question the motivations of the scientists who chose to continue working for the MRD after the war. Those who left may well have been concerned about the offensive aspects of the MRD's work and unwilling to continue the upside-down thinking required of microbiologists in weapons work. But there were attractions for those who stayed: such work fulfilled a sense of patriotic duty, maintained the excitement of wartime research, and was an opportunity to use the excellent facilities that the British government had built for the MRD. Finally, David Henderson created a unique and attractive working culture for scientists at the department. The MRD continued the wartime tradition of pulling in scientists from various disciplines to work together. Microbiologists worked with biochemists, physicists, meteorologists, instrument makers, experimental pathologists, and geneticists. Furthermore, scientists worked in groups, reporting to no one but the director (Henderson). Visitors to Porton (mainly Americans from Camp Detrick) tried in vain to create hierarchical organizational charts for the MRD, a point in which Henderson took some pride. In an important 1954 lecture to the Royal Society (London), Henderson quoted an American colleague who observed, "it may be said that everyone's work is directed by their conscience and Dr. Henderson."[33] Scientists' opportunities included groundbreaking laboratory work and adventurous field tests of live pathogens carried out on rubber rafts in the Caribbean.[34] In the late 1940s and early 1950s, the MRD was clearly a stimulating place for scientists because of the working culture and the national importance of bioweapons-related problems.

One of these factors was about to change, however. The successful British test of an atomic weapon in October 1952 meant that offensive *biological* weapons development began to slip in national priority. The Chiefs of Staff decided that the MRD (and its successor, the MRE, or Microbiological Research Establishment) should concentrate on developing defensive, protective technologies and long-term basic research on microorganisms (but not with an eye to making further weapons). The program would continue—the government's investment in new facilities after the war ensured that—but it would not have the cachet of the nuclear program. The scientists on the technical advisory committee grumbled that biological weapons development was being abandoned, and they were right. The Chiefs of Staff believed that their American collaborators would have a workable weapon in the near future in any case. For reasons of "economy and efficiency," the British program would now restrict itself to "experiments" on potential agents rather than "trials" of potential new weapons.[35] Historian Brian Balmer argues that this policy change meant a reconceptualization, not an abandonment, of "an ambitious research program" because "it was imperative for those involved in the area that the status of biological warfare be revived within the ambit of a defensively oriented policy regime."[36] In other words, David Henderson and his civilian scientists needed to find ways to keep their research projects running and preserve the unique culture of the MRD.

Henderson campaigned tirelessly for his unit with military and political officials and the nation's leading scientists. In a lecture delivered on April 1, 1954, to the scientists of the Royal Society, Henderson laid out the "why" and "how" of continuing research on BW. First, he characterized the MRD as a "defence research organization within the Ministry of Supply." Its goal during the war had been to "counteract" the "evil" of "deliberate dissemination of disease-producing organisms" by Britain's enemies. After the war, however, Henderson sought to recast the MRD and its new state-of-the-art facilities as a site of scientific knowledge production, open to "workers who have a legitimate scientific interest in our endeavors" who could make use of the "special facilities."[37] These facilities were the newest of Britain's microbiological research institutes, Henderson argued, not weapons factories. Yes, the agents studied were "unusual," as was the MRD scientists' focus on the "initiation of infection by routes other . . . than those accepted as natural portals of entry" (by this, he meant the focus on inhalational agents). And practical studies of how an enemy might attack with BW remained a "narrow channel of activity." Nonetheless, these practical studies had a "negative value" for the "advancement of knowledge," and Henderson made them sound more like an irritating requirement than a central focus of the MRD's work.[38] By spending the rest of his lecture discussing MRD research into the mechanisms of respiratory infection, the chemical basis of virulence, and the principles of immunity to anthrax and other diseases, Henderson steered his audience away from ideas of microorganisms as weapons and of the MRD as the site of offensive research.

The return to fundamental questions about disease pathogens

required reinitiating the domestication process for *B. anthracis.* Collecting new strains was a continual requirement in any case, because some did not survive or perform well in captivity. But if one wanted to study natural *B. anthracis,* the artificial strains of the laboratory were useless. The artificial method of "the test-tube culture of pathogens" had "outgrown its usefulness" to some areas of basic research; scientists needed to go back to the field to collect samples from diseased animals. As they had suspected, "the organisms as harvested had biological and chemical properties which differentiated them sharply from the same strain of *B. anthracis* grown in a wide variety of conditions *in vitro.*"[39] The manipulations of laboratory life had converted *B. anthracis* strains into artifacts, hypervirulent but unnatural microorganisms useful for weapons development but not for research on the natural history and life of the organism. Bringing new samples of first nature bacilli into the laboratory allowed scientists to begin the process of domestication all over again and to shape the microorganisms they had collected. This time, however, scientists could think right-side up, using experimental procedures such as studying the course of the disease *in vivo* (in living animal bodies).

The MRD scientists' actual experimental procedures differed little, however, in a practical sense, from what had been done during the war. Collecting new strains also represented an opportunity to tap the biological diversity outside the laboratory, to find new pathogenic strains from which to create vaccines, for example. They continued to rely on the piccolo, the apparatus that Henderson and Woods had developed for weapons work, to study basic questions about respiratory infectivity. They found that *B. anthracis* infected an animal only after the bacillus had been removed from the lung surface by scavenging phagocytes (immune system cells that "ate" foreign matter). They also further elucidated why natural strains of *B. anthracis* were not very infective by the inhalational route.[40] In the immunity experiments, researchers infected mice and guinea pigs with the newly collected natural strains of *B. anthracis,* took samples, and analyzed the samples and the animals from which they came "chemically and physiologically."[41] In the process, MRD scientists were instrumental in uncovering a fundamental property of *B. anthracis* that was the most profound change in the understanding of the bacillus since Koch's work in the 1870s: the chemical basis of *B. anthracis'* virulence, the secret of its killing power.

Anthrax investigators had long debated the method by which the bacillus killed its victims. In the nineteenth century, theories included the idea that the bacilli asphyxiated the animal through their great need for oxygen and that they caused morbid chemical derangements.[42] Twentieth-century theories increasingly focused on the fact that animals died of secondary shock (loss of circulating blood volume and resultant failure of organ systems). At the MRD, David Henderson paired Harry Smith (a young organic chemist) with James Keppie (a somewhat older veterinary microbiologist) and urged them to unravel the mystery of *B. anthracis'* killing power. Wartime events made it possible for them to do so. Ample opportunities to study secondary shock followed from numerous cases of battlefield injuries, and Smith and Keppie enlisted the aid of their colleague Joan Ross (a histopathologist who had discovered the path of *B. anthracis* through the lung's phagocytes) in understanding the cellular processes described by physicians who had treated patients with secondary shock. Anthrax seemed to kill its victims this way, too; perhaps the parallels would help them to understand how the bacilli were affecting the body.[43]

Also, the availability of streptomycin gave them a new tool with which to query *B. anthracis*. Smith and Keppie noticed that streptomycin prevented death in an anthrax-infected animal if it was given early in the course of the disease, before an irreversible "intoxication" had occurred. Later administration of the drug killed the bacilli in the animal's body, but the animal died anyway of acute renal failure and other organ damage. Why? Careful chemical examination of serum samples yielded the information that an unusual protein was present in the dying animals' blood. Smith and Keppie called it *lethal factor* because if they injected just this protein into guinea pigs and mice, it killed them after inducing the pathological changes characteristic of anthrax. Serology confirmed that lethal factor lived up to its name. If horses were hyperimmunized with anthrax spore vaccine (based on Pasteur's method), they did not die, but their blood contained another factor that kept guinea pigs and mice alive when they were injected with the lethal factor.[44] Thus, the Porton scientists concluded that *B. anthracis* killed its victims not by any direct action of the rapidly multiplying bacilli, but by the effects of the lethal factor, a chemical toxin it produced. The horse blood that had protected the guinea pigs contained something that counteracted the lethal factor—a *protective antigen* (first named by another Porton researcher, G. P. Gladstone in the 1940s).[45]

Henderson assigned the project of finding the protective antigen in the horses' blood to Smith and Keppie's colleagues Richard Strange and F. C. Belton. Strange and Belton, after a convoluted search for the right culture media and fractionation methods, figured out how to separate out and crudely purify a protein that protected rabbits against a subsequent challenge with the lethal Vollum M-36 strain of *B. anthracis*.[46] They isolated this protective antigen (PA) by taking apart the strain of *B. anthracis* from which Max Sterne had made his vaccine in South Africa in the 1930s (see chapter 3). Widely known as the Sterne strain, its more precise designation was 34F2, and it was a nonvirulent strain (although no one knew why at the time). Sterne's vaccine, along with Chinese and Soviet spore vaccines, would eventually replace the old Pasteur vaccine for use in livestock around the world.[47] These vaccine strains also became staples of laboratory research because they provided important clues to the inner workings of the bacillus.

The Secrets of Protection

By 1954, David Henderson could eagerly envision the synergistic effects of his researchers' work leading to a new vaccine because the toxin that contained the lethal factor also seemed to initiate immunity in test animals. Did it contain another PA? Or the same one that Gladstone had first named in 1946? Moreover, the combination of Smith and Keppie's in vivo research with the in vitro work of Strange and Belton could overcome the problems with artificial versus natural *B. anthracis:* "The whole scene becomes more provocative as the work of Smith and Keppie on the isolation of the lethal principle in anthrax infection advances, maybe to meet that of Strange and his colleagues on materials produced in the test-tube."[48] If *B. anthracis* could be deconstructed to its active principles, then perhaps scientists could finally overcome the problems inherent in using the laboratory bacillus for either protection or infection in the field. It is at this point that we must consider the interactions between Henderson's group and the scientists working at Camp Detrick in the United States, because they successfully probed many of *B. anthracis'* mysteries only after pooling their knowledge and expertise.

At the end of the war, American scientists also had a choice to make: continue working for the military with dual-use agents or return to civilian life. The head of the Air-Borne Division at Camp Detrick, Theodor Rosebury, had been one of Henderson's close collaborators and had designed the American cloud chamber apparatus with Henderson's guidance. Rosebury, who had been one of those alerting the Department of Defense to the danger of BW at the beginning of the U.S. participation in the war, did not hesitate to leave Camp Detrick after the war. Referring in his *Peace or Pestilence* to the idea of continuing BW research after WWII, Rosebury wrote, "What was once necessary has become doubtful, so that the smell of evil hangs over it inexorably."[49] Invited to join the Pugwash group of international scientists devoted to peace, Rosebury's continuing dedication to the cause negatively affected his career at crucial points. He never again worked on *Bacillus anthracis,* and he did not continue developing his expertise in aerosolization of microorganisms.[50]

The American program retained its focus on offensive as well as defensive BW research, but its staff numbers decreased significantly. From a peak of around three thousand employees at the end of the war, the Camp Detrick (later Fort Detrick) compound employed only three hundred to four hundred people by March 1946.[51] Nonetheless, a new facility for BW work was completed in 1950, the whole program expanded, and military opportunities abounded with the outbreak of fighting on the Korean Peninsula. The Department of Defense placed high priority on bacterial agents (with *B. anthracis* at the head of the list), including generating development and production schedules. In the midst of the Korean War, Chinese and Korean officials accused the United States' forces of using biological weapons against military and civilian personnel on the peninsula. Historians have concluded that there is a "long circumstantial trail of corroborative evidence that the United States experimented with biological weapons in Korea," an accusation that American officials have long denied.[52] Americans continued their tripartite collaborations with Canada and the UK throughout this period. While the British focused more on basic research problems, the Americans continued field trials of weapons and benign *simulants* (analogues) and worked to make large-scale production of BW agents feasible. *B. anthracis* was still the most stable and well developed of the potential agents available for weapons testing.[53]

The innovations in the British and American programs mostly fell into two categories: identifying viral agents and deconstructing bacterial ones. The 1950 destruction of Australia's overrun wild rabbit population from a deliberately introduced viral disease, myxomatosis, piqued the curiosity BW scientists. Myxomatosis was a cousin to smallpox; perhaps novel weapons could be developed from such viruses.[54] On the bacterial side, British and American scientists focused increasingly on identifying and isolating killing mechanisms and host immune responses. George G. Wright won the U.S. Army's Exceptional Civilian Service Award for developing a filtered vaccine for *B. anthracis* to be used on human populations. Reported in the *Science News Letter* as a "model for making other vaccines," Wright's method consisted of filtering *B. anthracis* cultures that had been grown on a special nutrient agar that contained no blood or serum. By spinning and filtering the bacteria, Wright caught the microorganisms' spore components in a filter while allowing the protective antigen to go through. While much remained to be learned about the protective antigen, one thing was known: it was the most promising substance scientists had yet found for inducing immunity to anthrax with no danger of causing illness (whole bacilli were needed to produce disease).[55]

Even so, vaccines against *B. anthracis* remained difficult to use. Wright and his colleagues had developed the vaccine from a mutant strain of the highly virulent Vollum; precipitated the PA with alum; then tested it on rabbits.[56] Finally, some six hundred personnel at Detrick volunteered to be the first people to receive the vaccine. With no major adverse effects reported, Wright then supplied the vaccine to Philip Brachman and his colleagues (one of whom was Herman Gold—see chapter 3). Brachman and his group had an idea for a field study of the vaccine: test it on wool factory workers, the only natural human population regularly exposed to inhalational anthrax. They conducted the trials in Northeastern factories in 1955–1957. The scientists emphasized in their published work that participation in the trials had been voluntary, and they were explicit that the vaccine used contained no whole spores (thus could not cause disease). Nonetheless, bad luck or coincidence struck when an outbreak of inhalational anthrax (most likely from infected goat hair) occurred at the largest mill, interrupting the study there. Contamination of the vaccine preparation with a few live spores could not be absolutely ruled out in the minds of many people (then and now). The eventual public scandal stopped large-scale clinical trials with this type of anthrax vaccine on civilians.[57]

The British and American studies of lethal factor and protective antigen came together in 1957–1958, when Porton's Richard Strange spent a working year at Fort Detrick with Curtis B. Thorne.[58] Strange and Thorne continued the work that Strange had begun with Belton, taking apart the most virulent strains of *B. anthracis* and isolating the very molecules that it used to infect and kill host cells. Their technique, an early version of one still used today, involved separating the proteins in the toxin by sending them through a piece of paper with a small electric current. The lighter proteins traveled farther down the paper than the heavier ones did, and the electric charge separated them from one another (fractionation of the toxin). As it turned out, the toxin that held the secret of *B. anthracis'* killing power also held the secret of immunity to it. It was composed of three components: the lethal factor, PA, and an *edema factor* that caused swelling in guinea pigs when a dilute solution was injected. Separated from each other and injected into test animals, each of the fractions caused minor or no problems. Only together did they create the pathogenic changes characteristic of anthrax. PA was the key to a successful vaccine (although it was clear even circa 1960 that it was not the only factor contributing to the development of immunity).[59] By understanding how *B. anthracis* infected a host cell, scientists could formulate ways to prevent this from happening and generate immunity against the bacillus in people and animals.

An important change in anthrax vaccine studies can be traced to the 1950s: vaccines being developed against the potential use of the biological weapon replaced the purpose of vaccinating against the agricultural disease. Early vaccines had targeted livestock and, potentially, people who worked directly with livestock (by this time an ever-decreasing percentage of the population). But after World War II, fear of *B. anthracis* as a biological weapon meant that scientists working on anthrax vaccines for human use increasingly tailored their vaccines to the weaponizable strains. Empirical techniques had led to the development of vaccines based on modified live spores (including the Pasteur and Sterne vaccines), but these were most suitable for livestock. Because they comprised living organisms, they could cause side effects and even some cases of the disease; these effects were unacceptable in human populations. The work of Wright, Strange, Thorne, and the other scientists from the weapons labs was a new approach: take apart a (virulent) strain of *B. anthracis,* make vaccines from its components, and then challenge them with the extremely virulent strains of *B. anthracis* developed for weapons use. The component types of vaccine had carried few, if any, of the objectionable pathogenic effects and risk of disease and death associated with live spore-based vaccines.

Clearly, scientists in the 1950s were working toward an ever-greater resolution in their examinations of *B. anthracis,* allowing them to divide the microorganism into components that met their goals without unpredictable effects. In the process, they necessarily parsed *B. anthracis* into forms that were not found on their own in nature. Component vaccines (or toxins, for more nefarious purposes) functioned as triggers of desired events: immunity and survival in the face of a host's bodily confrontation with *B. anthracis.* They retained their ties to the original strain of the bacillus from which they had come, however, for reasons that no one really understood until the next giant step in resolution in the late 1970s, to the subgenetic level of base pairs and codes (a topic of chapter 6). But the pursuit of disassembling *B. anthracis* to create manipulable tools increasingly became a major focus for defensive research in the 1960s. As Hannah Landecker has observed for cell cultures in the laboratory at this time, the "very possibility of such technologically mediated life forms was in itself seen as a result or finding of scientific work."[60] In research on *B. anthracis,* this process reinforced the illusion of control of the bacillus and created an infinitely flexible suite of products based on the natural raw material.

This becomes evident with a brief look at the Soviet biological warfare program during the cold war. The Soviet program dwarfed those of other nations, especially regarding *B. anthracis.* Information about the program (still quite incomplete) only began to emerge following the dissolution of the USSR and the defections of some former high-level Soviet BW scientists. According to one of them, Ken Alibek, the Soviet Union sponsored well-funded and well-regarded research, testing and manufacturing of *B. anthracis* and other agents. The Soviets stockpiled hundreds of tons of *B. anthracis* alone. In Stepnogorsk, northern Kazakhstan, the Biopreparat laboratory produced more *B. anthracis* spores than any other facility in the world: three hundred tons per year.[61] The Soviet system had also been the only one during the war to produce an anti-anthrax vaccine for humans. By 1953, Soviet scientists had isolated STI-1, a new strain similar to the Sterne vaccine strain from which they made a licensed human vaccine.[62] Thus, the USSR was the only nation in the immediate postwar period to have the offensive and defensive capabilities necessary to using *B. anthracis* as a weapon. The Soviet system would continue to innovate in both areas, especially after the influence of Lysenkoism waned. Lysenkoism as a scientific policy in the USSR deprecated modern genetics as a Western fallacy, and this restricted biological research for over twenty years after the war. Beginning in the 1970s, however, the Soviet BW system paid careful attention to the ever-increasing knowledge about genetics. Not a year after signing the 1972 Biological Weapons Convention (which forbade offensive BW programs), Soviet officials approved the creation of Biopreparat, where innovative BW research on anthrax hid behind commercial vaccine research, development, and production.[63]

The massive Soviet BW program (if the defectors' reports are accurate) dramatically altered the basic properties of *B. anthracis* organisms in search of novel strains for weaponization. Soviet scientists at Biopreparat and other institutes manipulated the genetics of already virulent *B. anthracis* strains to incorporate antibiotic resistance and the capability to evade or alter the victim's immune response—while retaining the organism's exceptional killing capacity. The plasmid coding for the lethal factors of the toxin proved mobile and transferable to other strains of *B. anthracis* or even to other species of bacteria. In the 1980s, Soviet scientists led by Igor Domaradsky were able to transfer tetracycline-resistant genes from a strain of *Bacillus thuringiensis* to *Bacillus anthracis* by mixing and cultivating them together. The few *B. anthracis* cells from the mixed culture that survived a dousing with tetracycline had incorporated genetic material from the resistant *B. thuringiensis* and would serve as the foundation for a new strain of virulent, antibiotic-resistant *B. anthracis*.[64]

Other scientists worked in the new model of disassembling strains of *B. anthracis* to create anti-anthrax immunoglobulin and enhancements to vaccines. (One must assume that the radically altered strains to be developed as weapons were a basis for these investigations.)[65] Their manipulations revolved around "curing" the strains of their virulence while maintaining the ability to stimulate the victim's immune system. These derivations of *B. anthracis* and immune proteins against it would be necessary to protect military and civilian populations and domesticated animals, should the weaponized strains be deployed by either side. Of course, the Soviets were not the only state-sponsored program to deconstruct *B. anthracis* in the interests of creating defensive vaccines. Changes in political policies in Britain and the United States led to an increased emphasis (publicly declared, at least) on defensive research and development. This change in policy had meant that the Porton Down production facility would be retooled to produce vaccines against anthrax (and other agents). In the United States, the Defense Department transferred Fort Detrick's facilities to the Office of the Surgeon General. In 1969, the former BW units were consolidated under the purpose of developing defenses against BW threats and renamed the U.S. Army Medical Research Institute for Infectious Diseases (USAMRIID). These changes did not kick *B. anthracis* out of the laboratory household; both the American and British units retained virulent strains and the facilities needed to work with them.

Indeed, they were always looking for new strains. *B. anthracis* has long been recognized as "breeding true," or in more recent parlance, it propagates clonally; the offspring resembles the parent. *B. anthracis* naturally mutates slowly, but any mutations that occur get passed on to daughter cells. Over long periods, *B. anthracis* has remained remarkably unchanging (as has its close cousin, *B. cereus*). Scientists have applied the techniques of their times to tell one strain of *B. anthracis* from another. In the 1800s, colony morphology (shape, texture, color, etc.) distinguished one strain from another. Scientists attempted to use serology, in which the immune cells of vaccinated animals would clump with bacteria, in the early to mid-1900s to identify strains (although it did not work very well). At the same time, nutritional and chemical testing of different bacilli began to be specific enough to tell *B. anthracis* from *B. cereus* and, they hoped, even to distinguish strains within each species.[66]

Beginning with the development of genetics and molecular biology in the 1950s, scientists began to understand and transform *B. anthracis* genetically. They mixed the virulent *B. anthracis* (*B. anthracis* was still defined by its ability to kill) with a benign *Bacillus* strain in the bodies of mice and in flasks of broth and found that they could create something new: a nonkilling strain of *B. anthracis* that had components "different from that of any member of the *Bacillus* group studied so far."[67] Thus scientists now had at least four ways to create new strains of *B. anthracis:* heat it, chemically alter it, passage it through animals, or mix and recombine it with near relatives. Pasteur's laboratory group had created their vaccines by heating and chemical alteration; Sterne and his colleagues had passaged their strain through animals' bodies. Thus, vaccines were one important product of this search for strains. Today, scientists still seek new strains of *B. anthracis* (now defined by their genetic makeup, rather than the crude measure of their ability to kill) as the raw material for producing new types of vaccines.

But what were these processes—heat, passaging, and the rest—really doing to the bacillus? And who decided when a strain was something "new," "different" from the rest? In 1958, an international group of bacteriologists decided on and adopted an International Code of Nomenclature of Bacteria and Viruses in which they defined the ideal "type culture" of each species. This essentially standardized the basic identity of *Bacillus anthracis, Bacillus cereus,* and the other members of the family. The type cultures were defined as "strains typifying the established species," and they would be used as the baseline for comparing new strains.[68] Thus, in an important sense, the modern genus *Bacillus* and its family tree (as we know it) date back only to 1958 (and have been constantly argued over since).[69] The *Bacillus anthracis* type culture should have been that of Koch, according to scientific tradition; but the descendants of Koch's original cultures were destroyed in Berlin in 1945 and apparently were not preserved anywhere else. Instead, a joint British-American commission proposed that Porton Down's Vollum strain of *B. anthracis* become the type culture.

Indeed, *B. anthracis* has come to be widely represented by the Ames and Vollum strains—two of the virulent strains isolated from sick cattle that have become well-domesticated to the laboratory.[70] The identification, maintenance, and supply of *B. anthracis* strains became a business as well as a professional courtesy among laboratories around the world. In the United States, for example, the American Type Culture Collection (ATCC) maintained stocks of various *B. anthracis* strains from at least the 1920s onward. Confident that the organisms were under adequate control by scientists and governmental agencies, the ATCC has been a kind of general store for laboratories needing stocks to begin or replace their own cultures. In the process, choices of type culture stocks kept on hand at strain banks like the ATCC has helped to maintain the identities of the standardized strains of *B. anthracis*, most of which were the artificial products of years of laboratory manipulation.[71]

By 1960, these manipulations had identified toxins and begun disassembling the capsule of virulent *B. anthracis* strains. Along with lethal factor, edema factor, and PA, the presence of a protective capsule around the bacillus seemed to add to its virulence.[72] These components of the bacilli—the toxins and the capsule—provided the material focus for research, defined the strains of *B. anthracis*, and were the basis for vaccine production. By the 1960s, scientists knew that the presence or absence of one of the toxin components or capsule affected the ability of the strain to kill its victim; likewise, PA was necessary to stimulate the victim's immune system. Only with the developments of plasmid genetics in the 1970s did *B. anthracis* finally reveal the molecular basis for its secrets of killing and protection.[73]

In 1983, Perry Mikesell, Bruce E. Ivins, and their colleagues at USAMRIID first described toxin production in *B. anthracis* as dependent on the presence of a plasmid, a circular segment of DNA. Plasmids, unlike the chromosomal DNA of the bacillus, can be exchanged from one strain to another. Their presence also indicates that they were acquired by *B. anthracis* from another virulent bacillus sometime during its evolutionary history. Fully virulent *B. anthracis* strains contained two plasmids: pXo1 and pXo2. The first plasmid encoded for the production of lethal factor, edema factor, and PA, and the second formed the protective capsule (the Sterne vaccine strain lacked pXo2). Mikesell and Ivins coined the term *cured* to mean a strain that had lost one or more of its virulence plasmids, and they cured strains by heating them sequentially. This method closely resembled that used by Pasteur and his colleagues in 1881. Indeed, two "Pasteur" vaccine strains examined by the USAMRIID group contained no pXo1 plasmids. A century after the dramatic experiment at Pouilly-le-Fort, scientists had elucidated a mechanism that explains why Pasteur's vaccine worked. No plasmid, no toxin, and thus no virulence.[74]

By the early 1980s, then, scientists could provide molecular mechanisms for why the Sterne and Pasteur vaccine strains did not cause disease in most vaccinated animals; could distinguish between strains of *B. anthracis* based on the presence or absence of plasmids; and could speculate about new techniques for curing virulent strains and making vaccines. At about this time, USAMRIID sent a letter to a network of veterinary diagnostic labs, university research centers, and other institutes around the nation asking for any new strains of *B. anthracis* that they had collected. Among the strains sent to Fort Detrick was one they later named Ames because of its presumed origin at the National Veterinary Diagnostic Laboratory in Ames, Iowa. Ames became a favorite at USAMRIID, but many of the new strains enhanced the scientists' base of material for research. They applied all sorts of new techniques to the bacilli: immunofluorescence studies, bacteriophage typing, and chromosomal DNA sequencing to look for differences and homologies in the genetics of the bacilli.[75]

Many questions remained, however. Because *B. anthracis* mutated slowly, different strains' chromosomal DNA proved highly similar. It took weeks and months for even the most sophisticated laboratories to locate the one or two differences in the genomes—and the differences were extremely small. Nonetheless, the ability to distinguish between strains became very important in the last quarter of the twentieth century, as anthrax reappeared in various places around the world. Strain analysis could help to answer an important question: Was a particular anthrax outbreak caused naturally, or had someone deliberately spread a laboratory strain? Beginning in about 1970, several factors increased the possibility of a deliberate deployment of *Bacillus anthracis.*

THE POOR MAN'S ATOMIC BOMB

As the major state-sponsored offensive programs wound down in the 1970s (the United States) and the 1990s (the USSR), the resources and expertise and stockpiled organisms that had belonged to them dispersed in various ways. In the early 1970s, 2,200 American scientists and technicians with BW experience became redundant as the Department of Defense dismantled its program and destroyed its stockpiled microorganisms. The Soviets looked on in disbelief. "We didn't believe a word of Nixon's [1969] announcement," Ken Alibek later recalled. "We thought the Americans were only wrapping a thicker cloak around their activities."[76] But within twenty years, the Soviet scientists too found themselves out of work as the USSR disintegrated, facilities came under the control of former republics, and Russia re-evaluated its expensive, risky, and illegal BW program. Among the consequences of these changes were a large number of people with biological weapons expertise looking (sometimes desperately) for work, and several facilities abandoned, left contaminated and unguarded.

At the same time, several characteristics of biological weapons portended their future use. They were "much cheaper to develop" than either chemical or nuclear capabilities, as Gradon Carter, former head of the MRE, explained. They were also "very flexible; no other method of war is capable of use on the same strategic, tactical and small scales to produce . . . effects which can be lethal or merely incapacitating, protracted or short-lived . . . [and] the opportunities for clandestine preparations are considerable."[77] And the fact that many weaponizable agents were dual use meant that weapons programs could be easily camouflaged as vaccine- or pharmaceutical-production institutes and facilities. Because many biological agents required some incubation time between the victim's exposure and the first signs of illness, it could be difficult to link the perpetrator to the crime. Under the best of circumstances, the problems inherent in establishing attribution to a perpetrator meant that biological agents allowed for the possibility of plausible deniability if the perpetrator so desired.[78] Finally, the necessary expertise existed in the redundant scientists from the now-defunct major BW programs. To these workers, certain biological weapons (including *B. anthracis*) were a well-known quantity by the 1970s, after almost forty years of development in laboratories and on proving grounds. Biological weapons were the ideal "poor man's atomic bomb," as the American Office of Technology Assessment put it in 1993: they were affordable, accessible, and available weapons of mass destruction, and they were an obvious choice over the other weapons of mass destruction (nuclear and chemical).[79]

The new biological weaponeers included the rising military powers of Asia, smaller states on multiple continents, rogue states such as Iraq and North Korea, and factions or groups interested in terrorism (such as the Japanese cult Aum Shinrikyo). Most, if not all, had ties to the older major state-sponsored BW programs. Lists of offensive biological weapons programs from the 1980s to 2000 (necessarily somewhat speculative, since these programs are highly secretive) include the nations of Israel, Cuba, North Korea, Iraq, Libya, Taiwan, China, Pakistan, India, and South Africa (during the apartheid years).[80] Most got started using resources from the former major state-sponsored programs. Iraq, Cuba, India, and Libya, for example, had hosted visits by Soviet scientists during which their Soviet counterparts shared techniques for genetically engineering bacteria and viruses.[81]

Anthrax played an especially important role in both Iraq's and South Africa's biological weapons programs. Nasser al-Hindawi, often described as the founder of Iraq's biological warfare program, was a microbiologist who had worked on *B. anthracis*. One of his successors, Baghdad University professor Huda Salih Mahdi Ammash, was nicknamed Mrs. Anthrax. Ammash had received her master's and Ph.D. degrees from academic institutions in the United States.[82] Iraqi laboratories certainly worked with *B. anthracis;* they had purchased twenty-one strains of *B. anthracis* from the ATCC in the 1980s. Some were sent to Baghdad University, where Ammash worked. Iraqi laboratories also applied to Porton Down to acquire *B. anthracis* strains, but Porton refused to send them.[83] In the course of the 2003 American-led military invasion of Iraq, it turned out that the Iraqi program did not hold stockpiles of weaponized *B. anthracis.* Mrs. Anthrax, taken into U.S. custody near the beginning of the war, was imprisoned but then deemed to be of no further use to U.S. intelligence-gathering efforts and released after two years of interrogation.[84]

South Africa's biological weapons program, known as Project Coast, had the horrific distinction of being an integral part of a racialized apartheid regime. P. W. Botha assumed the presidency of the minority white government in South Africa in 1978. Feeling threatened by the native African population's unrest in neighboring states such as Rhodesia (now Zimbabwe) and South West Africa (now Namibia), the Botha government began a chemical and biological weapons program in 1981. Biological weapons sometimes create strange alliances and realignments. The South African program declared itself to be a defensive effort against the USSR and China (who supported the nearby African liberation movements), and its major adviser on weapons of mass destruction was apparently Israel. *B. anthracis,* the Marburg and Ebola viruses, and a range of other agents composed the arsenal, and the program investigated agents that could sicken or sterilize people of African descent.[85] The head project officer, brigadier Dr. Wouter Basson, traveled around the world to scientific meetings and institutes, reportedly acquiring strains and vaccines and operations manuals, and creating a remarkable network of contacts.[86]

Anthrax had long been a problem in the nation's livestock, which had prompted Max Sterne and his colleagues to develop the vaccine strains in the 1930s at the Onderstepoort Veterinary Research Station near Pretoria (known in the 1980s as Onderstepoort Veterinary Institute, or OVI). Due to the history of work with anthrax vaccines there (see chapter 3), OVI was a natural site from which to obtain expertise and strains of *B. anthracis*. The role that OVI and individual scientists who worked there may have played in Project Coast remains unclear. Circumstantial evidence implicates OVI personnel as having tested biological agents. Scientists employed by OVI also worked for Roodeplaat Research Laboratories (RRL), a known site of biological weapons development. While on trial for war crimes in 1999, Wouter Basson declared that the majority of South Africa's scientific research facilities, universities, and hospitals were involved in aspects of biological weapons research. The OVI is likely to have done animal testing of particular strains and agent formulations for Project Coast.[87] As RRL veterinarian Schalk van Rensburg testified, the "products" of Project Coast included *B. anthracis* embedded in envelope gum and chocolates, and anthrax was supposed to have been used to murder Russian military advisers in Lusaka.[88]

Project Coast ended along with officially sanctioned apartheid in South Africa. Former members of Project Coast, including physicians, left the country to live and work in places such as Canada.[89] In 1994, Nelson Mandela, an opponent of apartheid policies and man of African descent, was elected president of South Africa. Fearing that the now-unemployed Wouter Basson would take his considerable knowledge about biological weapons elsewhere (to Libya, perhaps), American officials persuaded Mandela that Basson should be rehired as a scientist by the government despite his past crimes. The war-crimes tribunal convened after Mandela's election targeted Basson and other former Project Coast officials, who mostly escaped punishment. One of the more dramatic revelations was the Project Coast plan (never carried out) to damage Mandela's mental capacity with toxic agents while he was incarcerated during Botha's regime.[90] Such was the chilling reach of biological agents in South Africa during the 1980s. Fortunately, the formulations of *B. anthracis* developed by Project Coast scientists were apparently not deployed.

But the situation in the southern and central areas of the African continent raised an important question: If anthrax was naturally present in an area, how could a biological weapons attack be distinguished from a natural outbreak? South Africa's northern neighbor Zimbabwe (formerly Rhodesia until independence in 1979) provides an interesting case study. In 1967, Max Sterne (South African developer of the livestock vaccine) asserted that Rhodesia had one of the lowest incidences of animal anthrax in the world.[91] In 1979–1980, anthrax emerged suddenly, killing tremendous numbers of livestock, especially in the Tribal Trust lands—the areas inhabited by the Shona peoples and other ethnic Africans. The human population suffered a 500 percent increase in deaths due to anthrax.[92] Anthrax has remained *hyperendemic* (high incidence, with sporadic epidemics) in animals in Zimbabwe ever since. Now present in the soils, capable of infecting both wildlife and domesticated animals, anthrax will probably continue to be a serious problem for decades to come.[93]

Was it a coincidence that anthrax appeared so suddenly and dramatically during the Zimbabwean war for independence? As with South Africa, this revolution conducted by a black African majority aimed at toppling a white-minority apartheid regime. The political and military situations were quite complex and chaotic, with guerilla armies taking control of the country in 1979. Even this chaos, however, does not explain the unusual pattern of anthrax's spread around Zimbabwe. As physician Meryl Nass has pointed out, the rapid spread of the disease to new areas (natural anthrax, tied to the soil, does not usually do this) and the fact that it affected mainly Tribal Trust (ethnic African) livestock made the epidemic look suspiciously human caused, perhaps by supporters of the doomed white regime.[94] Public health officials in Zimbabwe during the epidemic attempted to explain the unusual pattern of spread by studying water supplies, the movements of vultures, and biting insect populations. These explanations were not particularly convincing; nor could the epidemic be attributed to the importation of hides, meat, or bones since there were very few industrial cases. Most of the people who were infected were ethnic Africans living in the regions of the livestock epidemic, handling infected animals, and eating infected meat.[95]

If the Zimbabwe epidemic *was* an episode of sabotage in conjunction with the revolution, then who did it, how, and why? The short answer is that no one really knows. Nass, however, has constructed a theory of such a sabotage operation. "Weighing all available evidence," she wrote, "a plausible explanation for the sudden peak of anthrax in the Tribal Trust Lands beginning in November, 1978, is that one or more units attached to the Rhodesian military may have air dropped anthrax spores in these territories. This action would expose cattle to the disease through ingestion or inhalation (or both) of anthrax spores. Humans would have acquired the disease from meat or meat products."[96] Nass believed that the intended victims were the cattle; the people would be secondarily affected by the loss of their main food supply. Beginning in October 1978, the Rhodesian government carried out numerous bombing raids, some of which could have included *B. anthracis*–charged bombs.[97]

Nass's theory is just that, and it has been disputed by other authorities who believe that the outbreak was a natural one exacerbated by the breakdown in veterinary and human medical services in the Tribal Trust areas during the war. Regardless of whose ideas most closely reflected events on the ground in 1978–1980, Nass made an important point at the end of her analysis. Only by understanding the historical, genetic, and ecological characteristics of *B. anthracis* in Zimbabwe could we hope to understand how it became an anthrax district in the 1970s. "During the past 45 years, no allegation of biological warfare has undergone careful scientific analysis," Nass wrote. "There exists no generally accepted methodology to serve as a guide for the design of an investigation into the possible use of biological weapons . . . The time has come for a thorough inquiry" of the Zimbabwe outbreak.[98] An investigation, integrating several methodologies, would be the only way to settle the question of whether an outbreak was natural or caused by an artificial strain of the bacillus, dispersed by human action.

Crude anti-animal attacks and sabotage remain important in the global context. In many nations, the food supply is barely adequate in good times, let alone in wartime, so attacking animals with disease would be potentially quite effective against the civilian population. Anthrax could also be useful for sabotage against military forces that depend on animal power, as many small forces in guerilla-warfare situations do.[99] As so many commentators have pointed out, a wide variety of military groups would potentially have access to *B. anthracis* cultures and the expertise needed to convert them into crude weapons. Although it would be much easier and faster to make weapons out of bacilli that had already been domesticated to the laboratory, it would be possible to re-create the whole process, from soil bacilli to weaponized bacilli. Thus, a great deal of attention has been paid of late to the locations of anthrax districts and outbreaks of anthrax among wild as well as domesticated animals. These are the dangerous zones, where the least control can be maintained over *B. anthracis,* and they now exist on all continents except Antarctica.

RESISTANCE: THE "SACRIFICE ZONES"

In 1998, a curator at a Norwegian police museum was startled to find a stored glass tube containing two sugar cubes with a note attached that read, "a piece of sugar containing anthrax bacilli, found in the luggage of Baron Otto Carl von Rosen, when he was apprehended in Karasjok in January 1917, suspected of espionage and sabotage." Hastily, the curator transported the glass tube to the Norwegian Defense Microbiological Institute in Oslo. Both sugar cubes had holes drilled into them, with a tiny glass capillary tube containing a brown fluid buried inside one of the holes. The Norwegian team rang up their counterparts at Porton Down, who had the capacity to analyze extremely small amounts of fragile biological material, and the sugar cubes traveled to Britain. In the containment facility at Porton, the brown liquid emerged from the capillary tube for the first time, eighty years after it had been placed there. Not only did it contain *B. anthracis* spores, but they grew in enriched culture and contained both plasmids that coded for the cell capsule and the lethal toxin.[100] In other words, the spores that had waited quietly in the Trondheim police museum for eighty years were still capable of infecting people or animals exposed to them.

The episode was sobering confirmation of the ability of *B. anthracis* to resist time and environmental conditions. The spores remained potentially dangerous in any place they inhabited, even decades (or centuries) later. This environmental resistance is one of the qualities of *B. anthracis* that has kept it on the A list of biological weapons development programs around the world. Once deployed, *B. anthracis* would not spread from animal to animal as a contagious disease would; instead, after killing its first wave of victims, it would implant itself in the soil and environment of its new home. It would prove to be extremely difficult to kill or remove completely once established. At Gruinard Island, we see perhaps the most well-documented example of this fact.

Gruinard Island remained quarantined by the British government until 1998, almost fifty years after the *B. anthracis* field tests conducted there during the war. In a total of thirteen trials conducted in 1942–1943, 4-pound cluster component bombs and 30-pound bombs containing *B. anthracis* spores were dropped on the island. One of the bombs, dropped from an airplane, buried itself six feet deep in the peat as it detonated, thus sending spores directly into the soil. The spores had been created in the laboratory. Genetically, they were variations of the extremely virulent Vollum M-36 strain. Physically, some of the preparations had been processed to separate spores from each other, the better to create highly infective inhalational clouds of bacilli. The purpose of this "theater of proof" on Gruinard Island lay with comparing the casualty effectiveness of *B. anthracis* bombs with that of other weapons. By the time the last sheep carcass (teeming with bacilli) had been buried, "it was concluded that an anthrax bomb was several orders of magnitude more effective than the most potent chemical warfare agents known at that time."[101]

At the end of the war, the British government had a problem: fearing outbreaks of anthrax and discovery of the secret bombing trials, they could not return the island to its owners or open it to the public. Still, rumors among the local people could not be contained after a dead sheep washed up on the mainland in 1943, infecting a dog, cats, and livestock. Periodic anthrax outbreaks on the mainland over the next decades reinforced the idea that the island was the source (although fortunately no human cases were traced to it).[102] Hoping that the *B. anthracis* spores would die off over time, Porton scientists visited the island annually to take soil samples throughout the 1950s, '60s, and '70s. The contamination persisted—and so did the killing power of the spores.[103] Local people complained that the island remained a menace to their health and that the government's indifference amounted to treating them as "third world citizens."[104] Gruinard Island was a sobering lesson. As historians Peter Hammond and Gradon Carter have noted, if *B. anthracis*–charged weapons had been used in populated areas, it would still be a problem today. The decontamination process eventually used at Gruinard Island would have been all but impossible to carry out over large areas, and would have thus expanded *B. anthracis*' ecological range dramatically, creating new hyperendemic anthrax districts.[105]

As it was, British officials found decontamination at Gruinard to be difficult in both practical and political terms. A 1979 survey of the island by Porton scientists revealed that the load of *B. anthracis* spores had diminished enough to make decontamination a possibility. They also found the major clue to what forces might have diminished the spore concentrations over time: a "profuse" growth of "natural aerobic spore-forming soil microflora." Having to compete with these bacilli, and almost fifty years of sitting and waiting, had reduced the *B. anthracis* bacilli to concentrations as low as three spores per gram of soil, requiring the investigators to create new techniques for finding them.[106] Decontamination would be possible once the Porton group (Chemical Defence Establishment) decided which procedures and chemicals would be best used. One thing was certain: the effort would cost the government a great deal of money, a fact that no doubt kept the Gruinard decontamination project well down the priority list.

Indeed the effort may not have been undertaken when it was if it weren't for Operation Dark Harvest, a fascinating episode in the history of postwar BW protest. Almost forty years after the weapons trials on the island, a self-identified "team of microbiologists from two universities, aided by local people" claimed to have slipped ashore and removed contaminated soil from the island. The perpetrators left ten pounds of the soil in a carefully wrapped container at the gates of Porton Down (subsequent tests confirmed the presence of *B. anthracis*). Simultaneously, they sent a message to several British news outlets that demanded government action to decontaminate Gruinard Island. They would periodically deposit soil samples containing the "seeds of death" around the country, ensuring "the rapid loss of indifference of the government and the equally rapid education of the general public."[107] The event succeeded in drawing attention to the plight of Gruinard Island. Despite their claim of including scientists in their midst, the perpetrators were environmental activists who may have found willing collaborators in people on the mainland near Gruinard Island who were upset about the government's indifference.[108]

Regardless, their actions coincided with further studies at Porton on how to kill *B. anthracis* spores in the soil. Because the spores are notoriously resistant to being killed by disinfectants, the vaccine group at Porton turned their attention temporarily to deciding which chemicals would be most likely to succeed. They tested glutaraldehyde, formaldehyde, peracetic acid, and dodecyamine on the island; by 1983 they had determined that all but the dodecyamine worked.[109] The government hired an English company in 1986 to decontaminate the island by burning all vegetation, removing some soil, and infiltrating the contaminated parts of the island with a disinfectant. Formaldehyde in seawater was the cheapest, so it was administered through thirty miles of buried fenestrated pipe over about two weeks. The effort cost half a million pounds, using 280 tons of formaldehyde and 200 tons of seawater.[110] Afterwards, the island was allowed to rest and the seeded vegetation to regenerate for a year. Officials reached an agreement with a farmer on the mainland, who transported a flock of sheep to the island to graze during the summer months of

1987. The sheep were an assay of the decontamination effort's success; highly susceptible to anthrax, they would have been likely to sicken and die if any spores remained near the surface. Nothing happened, and in 1990, a government official ceremonially removed an off-limits sign from Gruinard's shore, declaring it safe again for civilian use and returning it to the descendants of the original owners.[111] Languard Holdings, the English company that had decontaminated the site, planned to market its new process for anthrax decontamination globally.[112]

With the decades of spore persistence on Gruinard Island, Porton Down's *B. anthracis* had proved its ability to move successfully from the laboratory into the environment. An artificial strain had been returned to the soil, and its physical persistence confirmed its ability to resist degradation by sun, wind, moisture, temperature, and other natural forces. This was an important first nature aspect of *B. anthracis* that, in the words of one historian, has "permanently altered the soil" at Gruinard.[113] Perhaps even more lasting than this physical legacy is British citizens' hardened mistrust of government officials and the Porton scientists when they declare the island to be safe. No matter how great the cleanup effort, Gruinard will remain the Isle of Death in memory and history.

On another island half a world away, the past is present. Vozrozhdeniye Island is small, uninhabited, and situated in the rapidly desiccating Aral Sea. Formerly a part of the USSR, it reverted to Kazakhstan and Uzbekistan when the Soviet Union dissolved in the early 1990s. The scene there today is eerie: in 1992, the Soviets hastily abandoned a long-used biological weapons field-testing site and a town that housed workers and their families (about 1,500 people) on the other side of the island (figure 5.1). The buildings and their contents—animal cages, test tubes, books and journals, even a fire truck—are still there under a heavy coat of dust, disturbed only by Kazakh scavengers.[114] The Soviets left another legacy: tons of anthrax spores brought to the island from Sverdlovsk, a secret BW facility in the mainland Ukraine, which were buried there in 1988. Today, this island is probably the single

Figure 5.1. Former biological weapons testing site, Vozrozhdeniye Island, Uzbekistan. In 1988, hundreds of tons of B. anthracis spores were dumped here, making this the most heavily contaminated area in the world. Judith Miller, "At Bleak Site, Killer Soviet Germs Survive, and Neighbors Worry," *New York Times* (June 2, 1999), A1. Courtesy Judith Miller / The New York Times / Redux. Used with permission.

most dangerously contaminated place on earth, potentially harboring not only *B. anthracis* but also the causative agents of plague, tularemia, brucellosis, typhus, Q fever, smallpox, and Venezuelan equine encephalitis (most of which are probably long gone). Many of the strains tested here had been drastically altered in the laboratory to be extremely virulent, resistant to antibiotics, or both; information is still emerging about the testing program.[115]

Testing of weapons on the island dates to 1936–1937, when the island was selected as a proving ground due to its extremely hot and dry climate and distance from any large city. With little activity in the ensuing years, Vozrozhdeniye Island remained quiet until 1954, when the postwar Soviet weapons-production program sought a field-testing site. The Ministry of Defense built three installations on the island: one for weapons testing; the Field Scientific Research Laboratory (PNIL) for analysis of defense and decontamination methods; and facilities for Military Unit 25484, which supported the scientific work.[116] There was also a town, containing shops, schools and homes, and an airport. The village sat on the north end of the island and the open-air weapons-testing range on the south end, because the prevailing direction of the wind was to the south. *B. anthracis* developed at laboratories in Sverdlovsk and at the Biopreparat center in Stepnogorsk regularly arrived for testing on Vozrozhdeniye's range. The scientists exposed horses, monkeys, sheep, and donkeys to it (along with smaller laboratory animals such as guinea pigs and mice). Tethered to poles that still stick up out of the ground, the animals breathed clouds of microbes, their bodies used as in vivo assay devices to see how quickly and fully the spores entered the lungs, penetrated the immune system, and caused the characteristic pathologies of anthrax. The spores included both the ordinarily pathogenic strains of the bacilli and specialized, hypervirulent strains developed for military use. Apparently there were laboratory accidents, including illnesses among people who spent time on the island and outbreaks in local animal populations.[117]

The island itself is, of course, heavily contaminated with *B. anthracis* spores from these various strains. So are areas of the Aral Sea that surrounds it. In 1988, Ministry of Defense officials increased the magnitude of the island's contamination by ordering the burial of several barrels of *B. anthracis* spores hastily removed from the Sverdlovsk facility (under threat of inspection for violations of the Biological Weapons Convention). Doused in bleach and thus supposedly inactivated, the spores within the barrels came from highly virulent strains. In 1991, with the dismantling of the USSR, PNIL scientists left the island, and the laboratories were closed. Other personnel followed as the island began the process of transfer to its new overseers, Uzbekistan and Kazakhstan, who inherited the massive biological contamination problem but could not afford to clean it up. In parallel with these major political changes, an equally momentous environmental change had become obvious on the Aral Sea. With its main source-rivers diverted for agriculture, the Aral Sea has been drying up since the 1960s, exposing more and more land and uncovering things that had been dumped under the water. In 1962, Vozrozhdeniye Island covered about 200 square kilometers; it had expanded to 2,000 square kilometers by 1990. Critically, a shallow zone has developed between the island and the mainland. As early as 2010, Uzbek residents may be able to walk onto Vozrozhdeniye Island, making it very difficult to secure the contaminated areas.[118] The presence of scavengers makes one suspect that the barrels of *B. anthracis* spores remained effectively unsecured for at least fifteen years. Former Soviet BW scientist Ken Alibek testified before members of the U.S. Congress in 2000 that "a determined organization or individual could obtain virulent strains of microorganisms from their natural reservoirs" or from artificial ones in places like Vozrozhdeniye Island.[119]

This concern prompted other nations to offer assistance to the island's new owners faced with the expensive decontamination effort. For example, the United States through the Department of Defense's Cooperative Threat Reduction program spent over $430 million between 1998 and 2007 to assist former Soviet republics with dismantling BW facilities, decontamination, and surveillance for infectious disease outbreaks (Kazakhstan and Uzbekistan included).[120] On the Vozrozhdeniye Island, bioweapons facilities were destroyed, and the barrels of buried *B. anthracis* spores were unearthed and disinfected in 2002. A joint American-Kazakh team found that, contrary to Soviet assurances, live and virulent spores still existed within the barrels. Finally, following the example of Gruinard Island, workers decontaminated portions of the soil itself.[121] Kazakh scientists worked to genetically analyze dozens of different strains of *B. anthracis* found in the region.[122] They are working against the calendar: if a land bridge opens between Vozrozhdeniye Island and the mainland in the next few years, people, rodents, birds, other animals, and insects could all potentially spread the lethal legacy of the island to the mainland.

While sponsoring decontamination in Central Asia, however,

the United States Department of Defense continued to tolerate its own contaminated indigenous field-testing site for biological weapons. Dugway Proving Ground in Utah was the site of numerous open-air weapons tests in the 1950s and 1960s, with agents such as *B. anthracis* and a host of others that have now set up endemic disease zones in local wildlife. Nor were the tests themselves limited to the area inside Dugway's boundaries. Other nearby public lands served as testing sites to determine how well the agents could spread among the native animal and plant populations. Areas controlled by the United States in Central America, the Far East, the Caribbean, and over the Pacific Ocean have also been used for biological agent tests. As the Environmental Protection Agency senior scientist Eileen Choffnes summarized it, these sites have become "disease reservoirs in perpetuity, essentially becoming 'national sacrifice zones.'"[123]

SIMULATING AN ATTACK

The resistance of *B. anthracis* spores to the effects of years of wind, sun, heat, and cold was certainly not unknown to scientists working with this agent in the mid-twentieth century. As we have seen, shepherds and other livestock raisers, along with animal and human healers, had long understood that anthrax tended to recur in "cursed fields." Therefore, they tested their lethal and resistant creations on the Gruinard Island and Vozrozhdeniye Island— uninhabited, secret, and expendable. But some tests could not easily be done even in the most secretive and totalitarian areas. Mass destruction by disease following deployment of, say, *B. anthracis* clouds in a populated area like a city could not be risked, yet scientists found it difficult to predict what the conditions in a crowded, complex urban landscape or a closed space (such as subway tunnels) would do to the dissemination and casualty effectiveness of their weaponized biological agents. The solution to this quandary was to use *simulants* as a stand-in for the deadly agent. For *B. anthracis*, *Bacillus subtilis variant niger* (otherwise known as *Bacillus globigii*) was the most commonly used simulant. A simulant did everything that the weaponized agent did except cause illness or death. Its dissemination could be measured using carefully placed air-sampling devices (just as on the testing grounds), and the geographic and meteorological effects of urban environments (or another human-occupied space likely to be targeted covertly) could be accurately assessed for its weaponized cousin.

Simulant tests have been conducted all over the world in cities bustling with millions of normal lives being lived. In 1950, a U.S. Navy ship dispersed *B. globigii* in the harbor near San Francisco's Golden Gate Bridge. The city where I work, Minneapolis, hosted a series of tests in August and September of 1952. While many details remain secret, an Army Chemical Corps report (written in 1953 and declassified in 1980) stated the objectives to be determining "street level dosage patterns," meteorological effects, "penetrations of the aerosol cloud into residences at various distances from the aerosol dispenser," and "any residual background or lingering effect of the cloud within the buildings." The buildings studied included homes, the Medical Arts Building on 9th Street (near the present Hennepin County Medical Center), and the Clinton Elementary School. The agent dispensed was zinc cadmium sulfide particles. This compound, in powdered form, simulated the aerosol distribution of bacilli in the same form; the "dosage area" was "unusually large."[124]

Political scientist Leonard A. Cole has argued that although the Chemical Corps briefed the mayor and other Minneapolis officials, the city's leaders were misled into thinking that the tests explored how to create a smoke screen to hide the city from a bombing attack. Citizens apparently were unaccommodating, vandalizing equipment and refusing to cooperate (one can imagine investigators were barred from entering some homes). Public deception of this sort continued in the next series of tests, conducted in St. Louis. There, the park commissioner apparently protested and may have been dismissed. But the experimenters concentrated their work in poor neighborhoods and stepped up the police support. The St. Louis neighborhoods tested were part of a "densely populated slum district . . . the Police Department . . . prepared to quell any disturbances resulting from the presence of the test crew in the area," according to the Chemical Corps report. As Cole dryly observed, the lesson of St. Louis versus Minneapolis was to "choose a slum where residents are less likely to be educated, inquisitive, or to question authority."[125]

Equally unnerving was the June 1966 spraying of *B. globigii* in the New York City subway, exposing more than a million people. *B. globigii* does not (and apparently did not at the time) contain the virulence plasmids; the army described it as "harmless" and "an excellent simulant for anthrax (*Bacillus anthracis*)."[126] By studying how the bacilli moved in the tunnels, officials could predict what might happen during an attack with a weaponized agent, either in the United States or in the subway tunnels of another nation. (This hearkened back to allegations of WWII-era German biological sabotage of the London Underground.)[127]

The development and use of simulants for open-air testing of populated areas depended on the laboratory work that had created artificial strains of agents for weaponization. As a 1986 report from the U.S. Department of Defense explained, "Living organisms can be altered to increase their toxicity or modify their effects on humans; substances which were too costly to consider can be readily produced in militarily useful quantities; and seemingly innocuous organisms . . . can be altered to produce deadly substances. . . The result is a manyfold increase in the number of candidate agents." Simulants were an essential component of defensive research against such "deadly substances," and they were even more difficult to make in the laboratory. "Developing simulants requires a great deal of work with toxic materials without any guarantee the simulant development will be successful." The simulant had to be both like and unlike the pathogen, and only extensive testing in the laboratory would determine the parallel properties of the two.[128] Despite these difficulties, other major BW programs during the cold war (including the UK and the USSR) developed them and tested them in large-scale outdoor trials—often without the knowledge of the people who lived and worked there.[129]

By the 1980s, scientists probing *B. anthracis* in these various ways had uncovered some of the secrets to its killing powers and had developed ways to protect humans and animals against it. With the development of human vaccines, *B. anthracis* could be used as a biological weapon without infecting one's own personnel. Casualty effectiveness could also be manipulated quite precisely, as the extensive field testing demonstrated. While scientists certainly remained puzzled by some aspects of the organism and how it interacted with a victim's body, they had a solid basic understanding of both *B. anthracis'* natural history and the artificial uses of it as an inhalational agent. *B. anthracis* had become a weapon. The major question that remained was this: Would it be used?

CHAPTER SIX

Detection and Verification

The Weapon and the Disease

✵ ✵ ✵

In looking at what's required for aerosolized, easily dispersed powder, this stuff came close to the criteria. This is something that was prepared [to be a weapon].

USAMRIID scientist John Ezzell, opening a letter containing powdered *Bacillus anthracis,* October 2001

Bob Stevens, a hearty British-born outdoors enthusiast, worked as an editor for tabloid publisher American Media in Boca Raton, Florida. Americans were still in shock from the September 11, 2001, terrorist attacks on the World Trade Center and the Pentagon as Stevens sat opening mail addressed to American Media one day late in September. Amid the usual communications about sightings of extraterrestrial beings, celebrities, and other potential topics for the tabloids, Stevens opened an envelope out of which fell a threatening note and a beige-colored powder. On October 4, Stevens entered a hospital because he was having trouble breathing. On October 5, he died, a victim of an extremely rare disease: inhalational anthrax.

Within a few days, one of Stevens's co-workers developed a case of cutaneous anthrax, and the building was closed and decontaminated. Then cutaneous anthrax struck two employees of New York City news outlets and a baby visiting his mother's workplace. Besides American Media, studios for NBC News and headquarters of the tabloid *National Enquirer* received letters containing *Bacillus anthracis.* So did the offices of U.S. Congressman Tom Daschle, where a staff member opened a letter postmarked October 9 and reported that it contained a fine, floating, off-white powder. Between October 21 and October 28, four postal workers in New Jersey and Washington, D.C., became ill with inhalational anthrax, and two died. A total of twenty-two people became ill, eleven with the cutaneous form of anthrax and eleven with the pulmonary form (five of whom died). All were exposed to the spores contained in the letters, either by opening the envelopes, working in a contaminated building, or coming into contact with a contaminated segment of the U.S. mail system.[1]

The letters and their contents ended up in a high-security laboratory at the U.S. Army Medical Research Institute of Infectious Diseases (USAMRIID), located in the old American home of biological weapons research, Fort Detrick, in Frederick, Maryland. Clad in protective gear, scientist John Ezzell stood before a specially ventilated hood as he opened one of the letters. What he saw stunned him. The powder in the letter floated freely in the air, seeming to defy gravity. He assumed that the powder was spores of *Bacillus anthracis,* as the accompanying letter stated. But although he had worked extensively with *B. anthracis,* he had never seen any preparation of the spores as fine as these. Staring at the floating powder, he understood the intent of the person who had placed the powder in the envelope: to make spores that would easily infect and potentially kill whoever inhaled them. The spore powder had been created to be a weapon.[2]

As a final chapter in this biography of anthrax, the following pages detail the use of *Bacillus anthracis* as a weapon and the results of this use—anthrax outbreaks and how people have responded to them. Up to this point, we have seen some of the historical processes necessary to shape the multiple identities of anthrax, recognizing that "anthrax" has meant different things at different times. Anthrax has variously been a disease of animals and woolsorters' disease in people working in wool factories. It has been a skin disease, a severe gastrointestinal infection in people eating contaminated meat, and a dreaded pulmonary affliction. It has also been the disease that results from exposure in "cursed fields" or infection by *Bacillus anthracis*. Finally, over the past fifty years, anthrax has taken on another identity: it is a manifestation of someone deliberately creating and using *Bacillus anthracis* as a weapon.

Weaponized in this case meant that a very virulent strain of the bacilli had been processed specifically to break up clumps of the spores, even down to separating individual spores from each other (allowing them to be light enough to float in the air). This process made these particles much more infective when the victim inhaled them because they could penetrate into the smallest recesses of the lungs. Once ill with anthrax from inhalational exposure to these spores, a victim had about a 90 percent chance of dying. Compared with anthrax acquired through gastrointestinal or skin exposure, the disease course was usually quicker, more severe, and much more likely to end in death. Woolsorters' disease in factory workers also resulted from inhaling *B. anthracis* spores, but by the early twenty-first century, it had become extremely rare. Thus, when cases of severe, whole-body anthrax did appear, the disease was often suspected to be the marker of biological weapons use. *Bacillus anthracis* was the weapon; anthrax was the result.

An incident of bioterrorism framed the disease in new ways. Natural outbreaks of disease could be blamed on bad luck or angry gods. But when the disease appeared as the result of malevolent intent, the disease sufferers became victims of human cruelty. Outbreaks due to bioterrorism resulted from human malice—they were what I will call *premeditated diseases*.[3] They exposed the worst in human nature and some of the deepest social fault lines. Bioterrorism incidents also usually led to renewed calls to "control" the organism that had been weaponized. Control could mean many things: keeping the organism out of the hands of people with malevolent intent, manipulating the bacilli to harness or neutralize their killing power, or deciding the locations in which the bacilli were allowed to live.

Premeditated is an unusual identity for a disease. The person or persons responsible for the outbreak—the weaponizer—had to possess both the intent and the time to plan the attack and some degree of technical skill. He or she must have intended to cause debility and death and must have engaged in specific actions to prepare the microbes or toxins used as weapons so that they would be maximally infective and deadly. In the quotation that began this chapter, John Ezzell commented on this combination of the weaponizer's skill—the powder came close to the "criteria" for an ideal preparation of inhalational *B. anthracis*—and the intent to make a weapon. As far as Ezzell knew, no one had prepared this type of *B. anthracis* in the United States since Richard Nixon announced the ban on offensive weapons work in 1969. The two nations with biological weapons programs at that time, the United States and the Soviet Union, both signed and ratified the Biological Weapons Convention (BWC) that prohibited the storage and construction of weaponized *B. anthracis*.

Yet anthrax outbreaks had appeared in both nations since then: the 2001 U.S. anthrax letters and the outbreak in 1979 (affecting both animals and people) around Sverdlovsk, a city of 1.2 million people 1,400 kilometers east of Moscow (see chapter 5). Were these outbreaks the results of exposure to weaponized bacilli? If so, then what was the source of the bacilli? Answering these two questions requires a two-tiered investigation first to determine whether the outbreak was the result of natural disease or of the presence of microorganisms built to be weapons. The latter case would trigger a second, criminal investigation to determine who was responsible for building the weapons and inflicting them on the victims. Although different in many ways, both the Soviet and the U.S. anthrax outbreaks led to this pattern of investigation.

First, the outbreaks became the subjects of scientific investigation to detect and verify the source of the disease. Detecting and verifying were necessary to determine that the spores were weaponized and to validate the anthrax outbreaks as acts of bioterrorism or biocriminality.[4] Detection involved analyzing the bodies of sick people (or animals) to name the disease and determine its distribution in the population. Investigators relied on "shoe-leather" detective work, tracking down and interviewing the people involved, then determining how the bacillus had traveled and how the disease had spread. They also investigated the medical evidence, reviewing samples of the causative organisms and the victims' body tissues. Next, the goal of verification was to identify the cause of the disease, its source, and, if possible, whether the outbreak was likely to have been an act of bioterrorism. The bacillus itself, taken apart and examined carefully, could tell experienced scientists which other bacilli it was related to, where it had come from, and how it had been cultivated and processed. Detection and verification thus framed the disease in a new way and gave it a new meaning—either it was a naturally occurring outbreak or a weapon at work. The events that followed the 1979 outbreak of anthrax near Sverdlovsk illustrate this.

HUNTING THE SOURCE OF THE BACILLI: SVERDLOVSK

In April 1979, a few days before people began to die, anthrax broke out in livestock pastured within 50 kilometers of Sverdlovsk. Soviet authorities initially blamed the human outbreak on infected meat from the animal outbreak.[5] Outbreaks due to infected meat had regularly occurred in other Soviet locations, as the medical literature made clear. Moreover, *B. anthracis* was ubiquitous in the soil in and around Sverdlovsk, and more than 150 small outbreaks of anthrax among the local livestock had been reported in this region between 1936 and 1968.[6] *B. anthracis* had become domesticated to the region, with spores living in the soil of pastures and even in the nooks and crannies of some of Sverdlovsk's buildings. Natural outbreaks in animals could thus be expected with some regularity, as local residents well knew.

But many in Sverdlovsk (such as the pathologist Faina Abramova, who saved tissue samples from the human victims' bodies) doubted this explanation from the beginning, suspecting that the government was trying to cover up some kind of experiment or accident. They feared that a secretive Sverdlovsk military facility, Compound 19, had been the source of the anthrax bacilli. The severity of the disease and signs in the victims' bodies after death pointed to inhalational anthrax from very virulent spores, not the less deadly gastrointestinal form that would have resulted from eating infected meat. The victims died of respiratory disease and sepsis, not abdominal infections. Gastrointestinal anthrax did not explain the overall pattern of illness and death, either. People became ill before meat from the sick animals could have gotten into the food chain. The Soviet government's story did not add up. But it was difficult to know for certain, even for the local physicians and pathologists who had access to the scientific literature. Other than some case reports of industrial victims (woolsorters' disease rarely occurred in the twentieth century), information about inhalational anthrax outbreaks outside the laboratory was scarce. Finally, the KGB confiscated most of the records and evidence during and immediately after the Sverdlovsk outbreak in 1979, making detection and verification difficult.

After the dissolution of the Soviet Union in 1989, articles began appearing in the Russian press about the theory that the Sverdlovsk outbreak was caused by a biological weapons leak.[7] Analysts outside the Soviet Union had also suspected that an accidental leak from Compound 19 had caused the anthrax outbreak.[8] Perhaps in response to these pressures, Russian president Boris Yeltsin ordered an internal investigation. In May 1992 the Moscow newspaper *Komsomolskaya Pravda* quoted him as saying, "the KGB admitted that our military developments were the cause" but offering no further information.[9] The year, 1992, was important for another reason: Ken Alibek, a highly placed Soviet bioweapons scientist, defected to the United States. Alibek told American authorities that the Sverdlovsk facility was one of four or five key production facilities for *Bacillus anthracis*. These facilities together could produce nearly five thousand tons of weaponized anthrax each year (circa 1992). The work of weaponizing was done under extreme secrecy, even at the price of safety.[10] The only way to verify the source of the outbreak would be to conduct a unique retrospective epidemiological investigation that combined interviews with the people involved and an independent analysis of the scientific data remaining from the outbreak.

That was exactly what U.S. molecular biologist and arms control expert Matthew Meselson had in mind. Only a few years after the events of 1979, Meselson had applied for a permit from the Soviet government to visit Sverdlovsk and study the outbreak. Although Meselson at first found the meat-borne outbreak explanation to be plausible, his request was denied. With the dissolution of the Soviet Union in 1989, the new climate of Russian political openness encouraged Meselson to reapply, and he received permission to bring a team of investigators to the Sverdlovsk area. As the Meselson team prepared to go (they departed in June 1992), Soviet weapons stockpiles and labs were being hastily disposed of and dismantled, and many scientists with weapons-development experience found themselves out of work and unable to feed their families. Reportedly, former bioweapons experts were now selling sausages in the street. Scientists interviewed by American journalists described how easy it would be to steal vials of anthrax and sell them on the black market.[11] Thus the Sverdlovsk investigation had the potential not only to expose weapons treaty violations by the former Soviet government, but also to find evidence of how extensive the Soviet anthrax program had been and what had happened to its scientists and the *B. anthracis* spores it had produced. Even with Yeltsin's public admission that the outbreak had been caused by "military activities," Meselson's investigation by "independent scientists," as he put it, could potentially yield important information.[12]

Sociologist Jeanne Guillemin, married to Meselson and a member of the investigative team, wrote an authoritative, thoughtful, and poignant account of the investigation in her book *Anthrax*.[13] The book and several published scientific articles described how Meselson's team, including veterinarian Martin Hugh-Jones, pathologist David Walker, vaccine expert Alexis Shelokov, and local academics and physicians combined two investigative approaches. Along with examining the tissue samples and slides saved by pathologist Faina Abramova, the team planned a classic epidemiological approach to piecing together what had happened during the outbreak. They located and interviewed survivors, plotted the locations of victims on a map of the area, and cross-checked this information with various types of records, including those hidden from the KGB by hospital director Margarita Ilyenko. They also relied on historical information about inhalational anthrax: the research in the 1950s at Porton Down on *B. anthracis* aerosols, and published accounts of the New England outbreak of woolsorters' disease in the mill workers. By painstakingly correlating the geography and timing of the outbreak, earlier published work, and the physical evidence that still existed in Sverdlovsk, Meselson's team marshaled convincing evidence supporting a hypothesis that an airborne dispersal of weaponized *Bacillus anthracis* had caused the deaths.

The investigators spent almost two years on painstaking research. Despite Yeltsin's 1992 admission, the researchers approached the verification question—whether the Sverdlovsk outbreak was a premeditated disease—with an open mind. Gastrointestinal anthrax was certainly not out of the question as the cause because (as they knew) anthrax did occur naturally in the area. However, the victims' relatives (interviewed by Guillemin) asserted that what happened in April 1979 had *not* been a natural outbreak; their relatives had been ill with pneumonia, not gastrointestinal problems. Once the victims became ill, they died within three to four days. And there seemed to be little correlation between illness and having eaten (or not eaten) meat products.[14] Based on their information, Guillemin and her colleagues mapped the daytime locations of over sixty victims on the day of the accident and found a geographic pattern that placed them in an ellipse-shaped corridor southeast of the military bacteriological laboratory (figure 6.1).

This distribution of cases strongly suggested that the *Bacillus* had been carried to its victims by the prevailing winds on April 2, 1979. To create the geographic pattern of the victims, wind would have to have been blowing northwesterly—quite unusual for Sverdlovsk. Checking meteorological data, Meselson's team found that the wind blew from the northwest in Sverdlovsk on only about 2 percent of the days of the year—but that in 1979, one of those days had been April 2.[15] Guillemin wrote that when she realized what the evidence was telling her, the hair rose on the back of

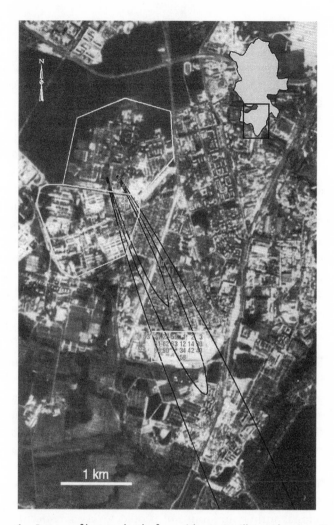

Figure 6.1. Pattern of human deaths from airborne Bacillus anthracis at Sverd-lovsk, 1979. Plotting the geographic locations and timing of the deaths and correlating them with the wind direction led to this oval-shaped estimate of the spreading cloud of bacilli. This pattern clearly pointed to a local military compound as the origin of the airborne bacilli. From Matthew Meselson, Jeanne Guillemin, Martin Hugh-Jones, Alexander Langmuir, Ilona Popova, Alexis Shelokov, and Olga Yampolskaya, "The Sverdlovsk Anthrax Outbreak of 1979," *Science* 266 (Nov. 18, 1994): 1204, figure 2. Reprinted with permission from AAAS.

her neck.[16] Only a concentrated airborne aerosol of *B. anthracis* released by the local military installation, Compound 19, could explain the evidence. A natural outbreak of anthrax did not usually sicken the respiratory system; nor would it have killed so many people in this geographic pattern. As Meselson later stated, "Bad meat doesn't travel in straight lines out to 50 kilometers south of the city."[17] The tissue samples clearly demonstrated that the victims had *B. anthracis* in their bodies and that they had suffered classic pathological signs of the disease. But if this had indeed been an inhalational exposure of weaponized *B. anthracis* from the military compound, why did another wave of cases occur about a month after the probable accident date (April 2)? Perhaps the spores had swirled around for several weeks, especially after the extensive street sweepings normally associated with the May 1 holiday festivities in Sverdlovsk.[18] But randomly swirling spores did not fit the geographic pattern of victims, and the Meselson team remained puzzled.

Meselson eventually found the answer while reviewing the historical literature, including work done in David W. Henderson's laboratory at Porton Down in 1956. Studying monkeys, Henderson's group found that spores could remain capable of causing infection for up to ninety days after inhalation and that the incubation time in the lungs varied with the inhaled dosage of spores.[19] Thus, the later Sverdlovsk victims could have harbored spores in their lungs for a period, becoming ill and dying a few days after the spores were activated. Meselson also searched for reports of inhalational anthrax in groups of people, a very rare occurrence. The most helpful report came from the vaccine trials and the concurrent outbreak of woolsorters' disease in the New England mill workers, published by Philip Brachman and his colleagues in 1960.[20] The epidemiological patterns and pathological data in the Sverdlovsk outbreak fit what the New England physicians had observed.

More and more, the epidemiological evidence pointed toward a premeditated disease—inhalational anthrax—rather than a natural outbreak of gastrointestinal disease. But verification also required that the investigators confirm the source of the bacilli and determine whether the release of the bacilli into the air had been an accident or intentional. If the former, then the outbreak was still a premeditated disease because the bacilli had been carefully prepared to be a weapon, but at least the weaponizers had not intended to unleash their weapon on an unsuspecting civilian population. If the release was intentional, the weaponizers were using their compatriots as human experimental subjects without their consent, and causing them severe harm. This would have been a serious accusation indeed.

Meselson's team could not definitively determine whether the release of the bacilli was accidental, because the KGB's official records on the incident (and the evidence they confiscated back in 1979) had been destroyed in December 1990. The top Russian official responsible for these records in 1992 was the Kremlin-based councilor of environment and health, who told the investigative team that he personally believed that the *Bacillus anthracis* release from Compound 19 was an accident, although he had not seen the contents of the destroyed file.[21] When the team again visited Russia in 1993, they interviewed a general who had intimate knowledge of the equipment and procedures within Compound 19. The details are eerily reminiscent of David Henderson and Theodor Rosebury's laboratories at Porton Down and Fort Detrick in the 1940s and 1950s: testing chambers (similar to Henderson's piccolo) in which two monkeys at a time could be exposed to aerosolized *B. anthracis,* hermetically sealed rooms with pressure gauges, and filtering to remove bacilli from any air exhausted by the testing chambers. Could one of these filters have failed, accidentally releasing bacilli from the top of the building? The American team thought this was a likely explanation; as Jeanne Guillemin put it, "Soviet technology has a far from stellar past," and the filter technology being used in Compound 19 was fifty years old. An accident seemed most likely.[22]

Finally, the investigators attempted to verify that the victims had died after exposure to bacilli that had come from Compound 19. This was easier said than done. First, bacilli could not be collected from the source. Not only was access to the Sverdlovsk military facility forbidden, but the Compound 19 test chambers had likely been dismantled and the premises disinfected by the time the investigators arrived in 1992. Bacilli could, however, possibly be isolated from the thirteen-year-old tissue samples and submitted to genetic analysis. If the bacilli were identified as Ames, Vollum, or some other type of strain developed specifically for weapons use, the Sverdlovsk outbreak could be assumed to have been caused by a weaponizable strain, most likely originating from Compound 19. The technology had its limitations, however. Although the American geneticists analyzing the tissue samples isolated four strains of *B. anthracis* that came from the same families as Ames and Vollum, their methods (circa 1996–1997) could not get any more specific than that. The tests did confirm that all the victims' bodies harbored a mixture of strains of *B. anthracis,* which would have been highly improbable with natural infection.[23]

The question about whether the Sverdlovsk outbreak was natural or unnatural lay at the heart of the Meselson team's major conclusions, published in an article in the journal *Science* in 1994.[24] To test their hypothesis that inhalational exposure to an aerosolized (therefore weaponized) form of *Bacillus anthracis* was the cause of the outbreak, Meselson's team used various types of evidence: climatic data, the spatial pattern of the deaths, and the timing of the outbreak in the local animals and people. By subjecting the Sverdlovsk data to epidemiological, historical, and genetic analysis, Meselson and his group demonstrated *B. anthracis* in victims' bodies and pieced together a plausible (even authoritative) narrative about the outbreak.[25] They had detected bacilli in victims' bodies that were extremely unlikely to have been present in the environment, but instead must have been cultivated in a biological weapons laboratory.

The investigation could not quite reach the point of verification, however— the cause of the disease, its source, and a confirmation that this disease outbreak had been an act of bioterrorism. Despite Boris Yeltsin's admission that the outbreak was due to "military activity," questions remained. What was the intent of the weaponizers? With the available data, the Meselson team could provide strong but not definitive evidence that a leak of weaponized bacilli from Compound 19 was the source of the disease. Continuing efforts by local investigators in Sverdlovsk (including testing soil samples for bacilli) provided no additional answers. As Jeanne Guillemin admitted, "Some aspects of the Sverdlovsk mystery may never be solved."[26]

But much was revealed by this investigation of an extremely rare and large outbreak of anthrax in humans. The data allowed estimates of how many spores an average person on the street would inhale, how many it would take to infect and kill a victim, and the likely fatality rate in the exposed population overall—the very parameters studied by U.S. and British researchers during and after World War II. Moreover, the Meselson team brought back some new information about how the old Soviet biological weapons program had been conducted in the region and details about the Soviet human vaccination program. And finally, even without complete verification, the investigation brought to light the deaths of innocent people and helped to confirm the existence of the Soviet biological weapons program despite the secrecy. As Jeanne Guillemin concluded, "The strongest implication of our work [is] that the cover-up of the outbreak's source, no matter how or why it was perpetuated, was the real crime in this case." For his part, Matthew Meselson felt strongly that pursuing the investigation was important on its own because, as he told a skeptical Kremlin official, "others [need] to know that if they violate their obligations, they will eventually be found out" and held accountable for breaching the BWC.[27]

Was the Sverdlovsk outbreak of anthrax premeditated? Yes, although with a caveat: the bacilli had indeed been weaponized, but with a different target population in mind. Although their accidental release was not premeditated, their creation was; they were merely transferred to an unintended population. Even unintended accidents, such as leaks from covert weapons production facilities, reflected a "later purpose," as veterinarian and team member Martin Hugh-Jones would put it.[28] (In other words, if a poisoner set the poisoned soup at the wrong place at the table and killed the wrong person, he or she was still guilty of murder.) Jeanne Guillemin concluded that Sverdlovsk was "like a tabletop model for future incidents against which we have to defend ourselves . . . the scenario can be transposed to an intentional fatal assault on a vulnerable population." Sverdlovsk was more than just "an account of an accidental emission," and Guillemin did not hesitate to assign full culpability for the premeditated action of creating a biological weapon with the intent to use it someday. In this case, the culpable parties numbered in the hundreds: military officers, Soviet officials who covered up the covert weapons program, the people who worked within it, and anyone else who knew about it.[29]

The outbreak of a premeditated disease reminds us of the power of these diseases to inspire fear and social disorder. In Sverdlovsk, much of this fear arose from the deadly nature of the disease itself and from historical experience with it. The sudden deaths of otherwise healthy people, the rumors of animal bodies piled in outlying areas, the hasty vaccination programs, and the seeming inability of government officials to end the outbreak and control the disease all shocked Sverdlovsk citizens. Russians had a long experience with and strong feeling toward anthrax; in a 1970 textbook coauthored by one of the generals interviewed by the Meselson team, anthrax was described as Russia's "self-styled national disease." In the 1800s, around 50,000 animals and 3,500 humans had died each year from anthrax. While that rate had fallen tremendously with vaccination campaigns after World War II, the more isolated rural Soviet republics such as Kazakhstan and Azerbaijan still experienced thousands of cases annually in animals and humans.[30] As everywhere, uncontrolled disease went hand-in-hand with deprivation and discouragement in the Soviet Union. Add to this the idea that the military had actually caused the disease and that government officials covered up the accident—and thus could not be trusted—and we can understand just how unnerving the anthrax outbreak must have been for Sverdlovsk residents.

VERIFYING ANTHRAX
The Deadly Letters

When the next human outbreak of premeditated anthrax occurred, in the United States in 2001, Americans in the affected areas experienced fear, uncertainty, and breakdowns in some social structures just as residents of Sverdlovsk had. More than three thousand people had died during the September 11, 2001, terrorist attacks, which used commercial airplanes carrying passengers. On that day, military airplanes whisked the president of the United States away to secret locations, the vice president activated a command center in a bunker under Washington, D.C., while fighter planes roared above the city, and the entire nation's commercial air system abruptly shut down. Within two weeks, Florida photo editor Bob Stevens died of anthrax; then the U.S. postal system was implicated as the source of his infection, and finally contaminated letters arrived at the offices of U.S. congressmen (or were intercepted on the way). Clearly, the anthrax victims had a premeditated disease, and this was another act of terrorism, which many feared was connected to the destruction of the airplanes and the attacks on the World Trade Center towers and the Pentagon.

For weeks, the U.S. postal system was seriously compromised, as two postal workers died of anthrax and all mail from the eastern part of the nation had to be isolated, irradiated, and checked. Ordinary citizens handled their mail with gloves or refused to handle it at all, fearing contamination with the deadly *B. anthracis* spores. In Washington, the Capitol and other federal office buildings near the fire-blackened Pentagon were closed for inspection and decontamination. The state of Florida, where several of the airplane hijackers had lived and where Bob Stevens was exposed to anthrax, was in an "acute frenzy," as bioterrorism expert Leonard A. Cole put it, and the frenzy spread in the coming weeks along with the reports of anthrax cases in other states.[31]

The Federal Bureau of Investigation (FBI), aided by the U.S.

Postal Inspection Service and eventually by scientific experts on *B. anthracis,* began an exhaustive investigation (dubbed Amerithrax) into the pathways traveled by the letters, their possible origins, and the nature of the spores within them. FBI-directed academic scientists and army scientists from over a dozen laboratories scrutinized the physical (especially genetic) characteristics of the bacillus that had been inside the letters. Genetic analysis of microbes had advanced since the Sverdlovsk investigation, and these scientists had more tools with which to analyze and identify the bacilli (and thus complete detection and verification). Still, it was no easy task. Identifying the individual strains of the bacilli and their exact origin would require over eight years, the time of hundreds of investigators and experts, a lot of money, and the rapid development of even more innovative genetic techniques for analyzing the identity of the *B. anthracis* strains.[32]

Indeed, to verify that the bacilli that killed Bob Stevens and others were weaponized strains, and to find their source, the science had to grow in concert with and be shaped by the criminal investigation. During the Sverdlovsk investigation, the Meselson team had come close to accomplishing verification by combing historical, pathological, genetic, climatic, and epidemiological data. Between 2001 and 2010, the Amerithrax investigators likewise united two distinct methodologies: the classic FBI forensic investigation (collecting evidence, doing interviews, ruling out potential suspects, etc.) and the rapidly changing genetic analysis of the microbes. Together, these methods have formed a discipline that is still developing as of this writing: microbial forensics. The negotiations over what microbial forensics *is,* and how it works, are currently being carried on by scientists, policymakers, and criminal investigators through the auspices of the American Association for the Advancement of Science and the National Academy of Sciences. The first textbook in the field was published in 2005, containing everything from the specifics of genetic analysis (and other ways of determining microbial identity) to laboratory management and admissibility standards for scientific evidence in a court of law.[33]

According to one definition that has emerged from these discussions, microbial forensics is the "analysis and interpretation of physical evidence to determine relevance to events, people, places, tools, methods, processes, intentions, [and] plans" associated with a suspected act of bioterrorism. The goal for geneticists participating in a microbial forensics investigation is *attribution,* or definitively linking a sample of microorganisms from a victim to a specific source of those microorganisms, even down to the very flask from which they came.[34] There is another step beyond attribution, however, which is linking the source of microorganisms to the person who created and released them and thus who caused an outbreak of premeditated disease. This is the goal of the law enforcement officers investigating potential bioterrorism, and it brings an even finer-grained analysis to verification. The Sverdlovsk investigators, working retrospectively on thirteen-year-old evidence, could not accomplish this. The bacilli in the victims' bodies were highly likely to have been weaponized strains but could not be definitively identified with the techniques available and thus could not be traced to their specific source. Moreover, there were no supporting documents because official Soviet records from Sverdlovsk had been deliberately destroyed in December 1990, ostensibly to protect those responsible for the outbreak.[35] Perhaps culpability mattered less in the Sverdlovsk investigation, however, since such a large number of people knew about the weapons facility. The weaponizers, the careless or tired operators in Compound 19, and Soviet military and public health officials all bore responsibility for the creation and escape of deadly inhalational bacilli. In the 2001 letter attacks, on the other hand, culpability proved to be a central question.

The letter attacks appeared to be the premeditated work of a group or individual (bioterrorists or biocriminals). On September 18, five letters containing *B. anthracis* spores were mailed to journalists, and on October 9, two more letters were mailed, one each to congressmen Tom Daschle of South Dakota and Patrick Leahy of Vermont. Then the letters abruptly stopped. Most of the letters did not reach their intended targets but did succeed in infecting other people by spilling spores as they were pulled through mail-sorting machines and were ripped open by secretaries and mail staff members. Thousands of people were treated with preventive antibiotics (which are most useful right after the victim is exposed, often before any symptoms appear).[36] Given the small number of victims, the response set off by the deadly letters might seem excessive were it not for the extreme climate of fear that already prevailed in the United States in the autumn of 2001. The FBI and U.S. Department of Justice made the anthrax investigation a top priority, together with the investigations of the airplane terrorists.

The rising numbers of anthrax victims and the discovery of the deadly letters sent shock waves around the world and triggered a global outbreak of anthrax hoaxes. In the United Kingdom, credible threats in London and Liverpool evacuated a stock exchange building and closed down post offices. Canadian workers in the parliament building complained of rashes after handling the mail, forcing parts of the building to be sealed and tested. White powder was mailed to the German chancellor, Gerhard Schröder, and to a newspaper office in Jerusalem. Australian prime minister John Howard called for calm amidst fifty-seven anthrax hoaxes in just over a week; authorities warned that hoaxers would be subject to up to ten years in prison if convicted.[37] In the United States, federal officials supported state and local authorities eager to crack down on hoaxers after dealing with more than two thousand anthrax scares between September 11 and October 15, 2001.[38] (This problem continued for years. One British postal worker, who "had the idea from the U.S. when anthrax had been sent through the post to various people in that country," sent more than 150 hoax letters and packages between 2003 and 2007.)[39]

The widespread concern over the deadly American letters reflected the potency and unique properties of the *B. anthracis* spores sent in the envelopes. Where had the spores come from—a terrorist group, or old Soviet stocks? Had the same perpetrator sent the letters, even though they contained two different formulations of spores? The first five letters held more crude preparations of spores that could clump together, but the two letters sent later to the congressmen contained the finer, "floating" spores that John Ezzell described at the beginning of this chapter. The finer the particles, the more specialized processing they had undergone—and the more likely they were to be inhaled, penetrate the body's defenses deep within the lungs, begin multiplying, and begin producing the deadly toxins that would kill their victims. (This is why antibiotics worked only within a day or two of infection, before the bacilli had a chance to flood the victim's body with toxins.) To make particles this fine, the perpetrator or perpetrators had the expertise to culture the bacilli, mill the spores, and dry them with additives that neutralized the electrostatic charges between the particles, thus making them float and increasing their aerosol spread.[40]

At first, officials from the administration of President George W. Bush strenuously asserted that the spores must have come from a covert offensive weapons program in Iraq, perhaps from old Soviet strains of hypervirulent *B. anthracis*.[41] Others suspected terrorists, perhaps the same group that had hijacked the airplanes on September 11. Some of the letters, including the one sent to Senator Daschle, said "Death to America! Allah is great!" suggesting that the perpetrator had religious motivations (figure 6.2). It seemed unlikely that the spores could have been domestically grown and processed, because few living American scientists had ever made an inhalational formulation of *Bacillus anthracis.*

William Patrick, retired from the USAMRIID, was one of the few who had worked in the old offensive weapons program at Fort Detrick and who remembered how to process bacilli into powder. The FBI eventually questioned Patrick and the other veterans of the defunct offensive weapons program, and searched their homes and cars, but turned up no evidence that any of these scientists had been involved.[42] The perpetrators had either gotten help from abroad (perhaps from veterans of the Soviet weapons program) or had figured out the process, acquired the necessary equipment, and done the work themselves.

Scientists cooperating with the investigation quickly deter-

Figure 6.2. An anthrax letter, 2001. Photo of an envelope containing *Bacillus anthracis* spores, addressed to U.S. Senator Tom Daschle. Courtesy Federal Bureau of Investigation, United States Department of Justice.

mined that the spores in all the letters had come from the same original growth culture, which supported the idea that the same perpetrator or perpetrators had mailed all the letters. By November 2001, geneticists determined that the spores in the deadly letters came from the Ames strains of *Bacillus anthracis,* which had originally been cultivated at USAMRIID in Frederick, Maryland. More and more, the deadly letters looked like the work of someone who was a scientific insider, someone with access to American-made strains of bacilli and the equipment to process it into a weapon.

Vaccines and Murder Weapons

Detection was accomplished: the spores were a very virulent strain favored by bioweapons researchers, and some of the powder was highly processed to make it very infectious. The creation of these spores had definitely been a premeditated act. Verifying who had created the spores and sent the letters, however, proved to be a difficult process. Ames strains of *B. anthracis* existed almost entirely in high-security laboratories such as the Dugway weapons testing center in Utah and the federal labs (used by both military and civilian contractors) that made and tested vaccines against anthrax.[43] Many of these were the defensive programs that had succeeded the old weapons development programs of the cold war era: the facilities at Porton Down and what remained of the old Soviet (now Russian) vaccine labs and production facilities. The U.S. vaccine development program was based on the old Fort Detrick compound, in USAMRIID. Other than that, cultures of Ames strains were stored only in a few university laboratories: Martin Hugh-Jones (Louisiana State University) and Paul Keim (Northern Arizona State University) kept the nation's largest collections of *B. anthracis* strains from around the world. Iowa State University's veterinary college held about one hundred strains of animal anthrax that faculty members had collected since the 1920s. (The latter collection was deliberately destroyed by the Iowa State faculty on October 12, 2001, with the permission of the FBI and CDC, because they felt it could not be secured against theft.)[44] The original source of what the FBI was calling the "murder weapon," the bacilli, had to have been one of these places—and the weaponizer may have been one of the scientists working there.

Even so, almost all the scientists recruited to help with the Amerithrax investigation were based in one of these facilities. The FBI had little choice in this matter; they needed the scientific expertise to conduct the microbial part of the forensic investigation. Right away, FBI agents had moved to secure the cooperation of scientists at USAMRIID and investigated the weapons testing center at Dugway Proving Ground in Utah after scientists there admitted to making small amounts of powdered *B. globigii* (*B. anthracis'* harmless cousin) to test protective equipment. (William Patrick supervised this process.)[45] Eventually, FBI officials developed unprecedented partnerships with Keim's lab and the scientists at The Institute for Genomic Research (TIGR, a private contractor) to do the complex genetic analysis. These geneticists had experience with a wide variety of microorganisms and were leading the way in developing techniques to tell one strain, or even one substrain, from another.

The scientists divided the work. USAMRIID scientists opened the deadly letters and captured the spores within them for analysis. They also analyzed thousands of samples from hoax letters, just in case one of them contained more of the virulent spores. The Keim lab selected *B. anthracis* strains from its extensive collection (and from other laboratories) that were likely to be related to the spores in the letters. Genomic scientists at TIGR spearheaded the work of sequencing the genomes of the spores from the letters and around twenty other *B. anthracis* strains selected by Keim and his colleagues. They hoped to find a match between the strains in the mailed spores and one of the other strains of known bacilli from known laboratories. (Identifying the laboratory from which the spores had come would greatly narrow the search for the perpetrator.) This was easier said than done. *B. anthracis* strains (families) and substrains (brothers and sisters within the families) are remarkably genetically alike. The genome of *B. anthracis* contains 5.5 million base pairs (genetic units), and the researchers had to find minute differences, on the order of only a few base pairs, to know whether two spores came from the same strain or substrain. It was slow, labor-intensive work; the geneticists were able to sequence and compare only small snippets of 250 base pairs at a time.[46] To help them, the Keim group tried to develop new techniques for genetic analysis on the fly that would be faster and more specific. Nonetheless, they estimated that it would take years to definitively identify the source cultures of the spores from the letters.[47]

Beyond the genetics, other scientists scrutinized minute details about the spores and their processing in search of clues. The extra-fine powder from the letters sent to congressmen Leahy and Daschle, for example, had been made using a particular kind of costly machine, a spray-dryer, which separated large clumps of spores into particles tiny enough to be inhaled deeply into the victim's lungs. This finding set off a hunt for the locations of spray dryers in the United States. Other laboratories used isotope analyses to ascertain sources of the nutrient broth in which the attack strain had been grown, and Dugway scientists tried to duplicate the perpetrator's method of creating the fine powder. Scientists even analyzed traces of the water in which the cultures of bacilli had once been suspended and the trace minerals that had gotten stuck to the spores. The FBI announced that the water analysis demonstrated that the bacilli had been cultured (grown) in water that came from somewhere in the eastern United States.

Then TIGR geneticists made a startling discovery in early 2002: their comparison of the genetic sequences established that the spores from the letters were practically identical to Ames strains kept at USAMRIID in Fort Detrick.[48] The bacilli may well have traveled around a bit (USAMRIID scientists commonly swapped strains with other scientists), but originally the spores had come from stock kept at USAMRIID. Under FBI orders, USAMRIID scientists submitted samples from their many stocks of Ames strains and began to swab the offices and hallways of their own workplace to determine whether the spores had been removed from the secured areas inside the lab rooms. They found Ames-type spores in many places, including on the desk of one scientist.[49] Although certainly suspicious, the implications of this contamination were far from clear. For one thing, the USAMRIID bacteriologists worked often with Ames spores to test their vaccine formulations. The contamination could have been accidental, accumulating over time, rather than evidence that someone had removed spores from the secured lab rooms.

Also, even though TIGR scientists had found one tiny distinguishing area on the genome of the attack sample, and therefore could narrow down its origin to USAMRIID, the DNA-sequencing techniques available at the time were not powerful enough to determine the exact identities of the sample's substrains. There were hundreds of flasks of *B. anthracis* at USAMRIID in the various secured laboratory rooms—which substrain, from which flask in which lab room, was identical to the spores in the letters? Knowing this would identify the people who had had access to the particular secure room and even the particular flask at USAMRIID from which the bacilli had been taken. Until the geneticists found more of the tiny genetic differences, or figured out faster techniques with higher resolution to sort through the 5.5 million base pairs and identify substrains of the bacilli, the scientific investigation of the spores was stalled. Like the Sverdlovsk investigation, verification of the identity and source of the bacilli was very difficult to accomplish.

Also like Sverdlovsk, the investigators seeking the identity of the bacillus and its creator did not operate in a social and cultural vacuum. Just as Soviet political ideology and military power had overshadowed the Meselson team's investigation, the Amerithrax investigation developed in a climate of fear and uncertainty in the wake of the 2001 terrorist attacks. It was also shaped by the previous decade of public notoriety regarding anthrax, developed in the context of political and military events in the United States. During the first Gulf War in 1991, 150,000 military personnel headed for Kuwait and Iraq were hastily vaccinated with one or two doses of an anti-anthrax vaccine. Some veterans experienced immediate and delayed effects that they associated with the vaccine. After the war, veterans' groups and their supporters continued to express skepticism about government vaccine programs. Nonetheless, amid increasing alarm about the possibility of a bioterror episode, Secretary of Defense William J. Cohen announced a new mandatory policy of vaccinating all military personnel against anthrax in December 1997.[50] This policy was very controversial, and some military personnel refused to take the vaccine as ordered. Investigative journalist Gary Matsumoto and physician Meryl Nass, among others, publicly described perceived problems with soldiers' informed consent and rumors of experimental vaccines being tested on military populations.[51] The military vaccination controversy dominated discussions about anthrax in the American popular press (and to some degree in Britain, Canada, and Europe as well) throughout the 1990s.

The process of creating, testing, and accepting vaccines was an important part of the response to the threat of premeditated anthrax outbreaks, including the letter attacks of 2001. After the 2000 election, President George W. Bush's persistent concerns (thought by some at the time to be exaggerated) about Iraq's possible biological weapons program and the breakdown of international weapons inspections encouraged public discussions about the risk of state-sponsored anthrax attacks. More moderate voices might have prevailed because, as Leonard Cole wrote in 1999, "during the last 100 years, the sum total of deaths in the United States known to have been caused by bioterrorism is zero."[52] Unfortunately, that changed with the death of Bob Stevens, the first victim of the 2001 anthrax letter attacks. Then in 2002, American military forces discovered a half-constructed laboratory at a site vacated by the terrorist group Al Qaeda, near Kandahar, Afghanistan. While there were no cultures of microorganisms present, the equipment being set up there would have been appropriate for producing weaponized *B. anthracis*.[53] The anthrax letters of 2001, and the implication that terrorist groups may have been developing biological weapons, supported the idea of keeping the anthrax vaccination program going despite the controversy.

The controversial history of military vaccination protocols also helps to explain why, in the popular arena, "anthrax" had become synonymous with bioterrorism by the turn of the twenty-first century in the United States and Europe. Vaccination had long been used by a tiny fraction of the population—mainly those at risk because of their occupations, such as veterinarians and wool factory workers. However, woolsorters' disease and cases of anthrax in farmers and veterinarians had become extremely rare and excited little controversy or publicity. The whole purpose of the controversial military vaccination program was to protect against a rather vague threat of bioterror activities using *B. anthracis* spores. The outbreak of human anthrax in Sverdlovsk, and the investigation that followed, certainly validated the idea that biological weapons caused anthrax.

After the 2001 attacks, public health officials offered antimicrobial drugs and vaccinations to people who had been potentially exposed to the deadly letters. Donald A. Henderson, at that time director of the Department of Health and Human Services' newly created Office of Public Health Preparedness, supported the use of the vaccine while cautioning that there was a risk of side effects. Of course, adequate supplies of the vaccine were a problem, given the tens of thousands of people who had been potentially exposed through contaminated mail.[54] *New Republic* journalist Wendy Orent even suggested that, given the insufficient supply of anti-anthrax vaccine for humans, the old Sterne livestock vaccine be considered for human use. "Given the circumstances—a series of anthrax attacks from a source with unknown capabilities, and a desperate shortage of the less-than-ideal AVA vaccine," she wrote, "human safety trials are worth a try." While her idea was not implemented, she cited at least one scientist who was willing to take the vaccine meant for cattle rather than not be vaccinated at all.[55]

The development and use of *B. anthracis* vaccines in the late twentieth-century United States depended on and shaped the identity of anthrax as a premeditated disease. Without the fear of attack with weaponized *B. anthracis,* there would have been little incentive to work on creating new vaccines (or improving existing ones). The scientists trying to develop new vaccine formulations worked with a serious purpose: to develop the means of saving military personnel's lives (even if some of those very personnel refused to take the vaccines). Yet the institute that housed research on vaccine formulations, USAMRIID, also seemed to be the source of the bacilli found in the letters. The murder weapon was closely related to tools used to test vaccines in experimental animals. Elucidating this relationship would verify the exact origin of the murder weapon (even to the single flask from which the bacilli had been taken, as it turned out), and perhaps even the person responsible.

MICROBIAL FORENSICS: STUDYING THE ZEBRA'S STRIPES

The Amerithrax investigators sought to definitively link a sample of microorganisms from a victim to a specific source, and then to link that source to the person who had created and released the microorganisms. They used the methods of microbial forensics, combining genetic analysis with the FBI's normal investigative techniques. Overall, the Amerithrax microbial forensic investigation took a little over eight years to complete.[56] By 2006, under the impatient eyes of FBI investigators, victims' families, and members of Congress, geneticists managed to develop and refine new experimental techniques that allowed them to determine the exact identity of the bacilli from the letters. To return to the genealogical analogy I used above, these techniques were able to detect tiny differences in the DNA of the bacilli to differentiate "brother" substrains from "cousins" and to pinpoint the location from which the bacilli had been taken. The Amerithrax investigation advanced the field of microbial genetics more quickly than it might have developed in ordinary times. But the scientists involved paid a price.

Working under the conditions of constant scrutiny was very stressful for both the geneticists and the vaccine researchers. The containment facilities staffed by USAMRIID scientists had steadily become sites of criminal investigation as well as of scientific research, even before the 2001 attacks, because of the tremendous number of anthrax hoaxes.[57] Responding to FBI requests to test suspicious powders kept a whole team of USAMRIID workers busy, with a particular facility set aside for criminal investigation work. After 2001, USAMRIID and other bacteriological facilities that housed potential biological agents were rapidly transformed by what one observer called the "sweeping policies to incorporate U.S. biological sciences into the campaign against bioterrorism."[58] Scientists had to become functionaries of criminal investigations. They were trained in the careful details of how to handle biological evidence so that their actions and conclusions could withstand potential challenges in courts of law (a key goal of microbial forensics).

The secrecy and vigilance of microbial forensics, however, made many scientists uncomfortable, and no doubt they found the work tedious and irritating. Conflicts arose between scientific culture and the culture of criminal investigation. For example, bacteriologists had long practiced a fairly open exchange of microorganism cultures between labs, a practice that became severely curtailed with new regulations in the 1990s.[59] Once the strain of *B. anthracis* used in the 2001 letter attacks was determined to be Ames, USAMRIID scientists were placed in the uncomfortable position of being simultaneously suspects and essential assistants in a criminal investigation. FBI agents mounted surveillance cameras in the labs and criticized many aspects of the scientists' procedures. They wanted the scientists to work faster, to strictly contain the bacilli within restricted areas, and to safeguard the lethal cultures to prevent theft or contamination. At the same time, the scientists endured polygraph tests and searches of their homes. John Ezzell complained bitterly about having to work in such an environment of suspicion. "You've worked so hard, and what have you gotten out of this? You get accused, and it's really disheartening. We're scientists," he emphasized. "We know there's a certain amount of necessity for all of this, but at the same time we're scientists—very dedicated, very loyal, very patriotic. And as hard as we've worked, to now be subjected to these kinds of observations is demeaning."[60] Yet despite (or perhaps because of) this pressure, the geneticists and bacteriologists who worked on the Amerithrax investigation did some of the most important and creative work of their careers under the glare of security cameras.

The Zebra's Stripes

As the geneticists told federal agents in 2001, the *B. anthracis* spores in the letters contained a genetic text that could be read and, they hoped, would provide the answers that the criminal investigators sought. Like a human fingerprint, or the stripes on a zebra, the genetic text was unique to each individual sample and could prove the origin of the *B. anthracis* powder and implicate the perpetrator. To do so, the geneticists had to solve several practical problems. First, the volume of the bacilli taken from the letters was quite small, and they needed to conserve it. So when they took their small samples from the bacilli's DNA to study them, they made millions of exact copies. This process left most of the original murder weapon physically intact while enabling hundreds of experiments on clones of its genetic code.[61] This was crucial because most of the experiments destroyed the cloned DNA, and multiple experimental runs were going on simultaneously in the labs of the FBI's multiple academic and corporate collaborators. They used up many clones of the original genetic snippets from the bacilli in the letters.

At Northern Arizona State University, Paul Keim's group of researchers took on a massive task for the Amerithrax investigation: to compare very small segments of the DNA from the murder weapon to the same small segments in the DNA of all the known existing cultures of Ames *B. anthracis*. As with USAMRIID, the geneticists in Keim's group found working as part of a criminal investigation to be uncomfortable. Even as they labored to assist the FBI, they wondered if they were also being considered potential suspects. Eventually, anyone who had worked extensively with *B. anthracis* could expect FBI interviews and inspections of their homes and workplaces.[62] Extra walls and locked doors went in at Northern Arizona State to secure Keim's collection of *B. anthracis* strains.[63] The members of Keim's lab worked under pressure in a fishbowl, purposefully and steadily hunting down tiny differences in the genetics of the bacilli.

These DNA comparisons did not just differentiate between substrains, but also brought to light characteristics of the human-caused processes that had shaped the spores. What types of manipulations may have been performed on them to make them more infective or more deadly? Had they been passaged through animal bodies? Had they been forced to mutate? Signs of artificial manipulations, if they could be found, could point to a particular laboratory, training institution, or scientist—very useful in a criminal investigation.[64] From the beginning of the investigation, the FBI placed a great deal of faith in these genetic techniques to identify the bacillus and the perpetrator—perhaps too much faith. Using these techniques to find the fingerprint of the pathogenic organism would not necessarily lead to the fingerprint of its human manipulators.

The "Forensic" Component of Microbial Forensics

The premeditated anthrax outbreak that resulted from the contaminated letters presented unusual challenges for the FBI, which was in charge of both overseeing the scientists and conducting the rest of the forensic investigation (the classic detective work). In the past, FBI investigations had been based in, and run by, the field offices closest to the incident. There, the agents had developed their networks of contacts, worked with the local police, and knew the social lay of the land. But incidents as geographically widespread and complex as the September 11 airplane attacks and the anthrax outbreak required wider coordination, and Thomas J. Pickard was given the unenviable task of heading both investigations simultaneously. The Amerithrax investigation was divided into two sections: Amerithrax 1, the classic detective work, headed by agent John "Jack" Hess, and Amerithrax 2, the scientific investigation of the bacilli, headed by David Lee Wilson. Hess, Wilson, and Van A. Harp, the director of the field office in Washington, D.C., formed the first team of FBI officials that determined the overall shape of the investigation.

Hess and the Amerithrax 1 investigators initially focused on analyzing the perpetrator's method (or modus operandi) and motivation. These characteristics were at the heart of determining a likely profile for the perpetrator. In this case, the perpetrator had been quite well organized, engaging in fairly technical and carefully planned actions and leaving behind few sloppy clues. The intent of the perpetrator seemed clear. Creating and deploying this type of *B. anthracis* preparation had been designed as an act of terror, to frighten the American population at large and cause premeditated anthrax. Yet some of the envelopes contained notes that advised the opener to "take penacillin [sic]," making the FBI's profilers speculate that the perpetrator did not want to kill anyone. The warning to take antibiotics may have represented the perpetrator's qualms about the anthrax letters, but his or her malicious intentions were obviously powerful enough to overcome those qualms.

The work of the geneticists, vaccine experts, and other scientists provided important information about the modus operandi—how the bacilli had been constructed and deployed in the letters. By the end of 2001, FBI agents had created several profiles of the perpetrator: a person who was a scientist or someone with access to laboratories, who held some kind of grudge, and who was a loner. Harp called the president of the American Society of Mi-

crobiologists, Abigail Salyers, and asked if she could think of anyone who fit the profile. "'Loner, loser and nerd' . . . that describes
at least half of our members," Salyers reportedly answered. The
profile was too vague to be helpful.[65]

Harp then sent a letter to all members of the society, asking if
they had any tips for the FBI. Since the origin of the bacilli had
been determined to be USAMRIID, Harp was convinced that the
answer lay within the American scientific community. In March
2002, he asserted in an interview that, "Contrary to what was
initially out there at the beginning of the investigation, this anthrax,
we do not believe, was made up in a garage or a bathtub . . . There
are only so many people, so many places that this can be done."[66]
The FBI sent out informational flyers and offered a reward for
information leading to the identity of the perpetrator. Within the
first six months of the investigation, agents had conducted about
five thousand interviews but had not yet found the culprit.

In August 2002, Attorney General John Ashcroft identified Steven Hatfill as "a person of interest."[67] The FBI pursued Hatfill
aggressively, searching his apartment, questioning his girlfriend
and her family, and finding that some of Hatfill's credentials
had been falsified. As Hatfill himself admitted, his background
would have been interesting to the investigators. He had been
serving in Zimbabwe's military in 1979, during the suspicious anthrax outbreak there, and he had connections in South Africa. He
had spent a short time working in USAMRIID's most high-level
biosecurity labs and had taken a week of training at Porton Down;
his family lived in Florida (where Bob Stevens had been infected);
and he gave public presentations and interviews on biological
weapons. Now he had the unwelcome attention of not only the
investigators but also the *New York Times* and other media outlets
that freely discussed the new "person of interest" in the investigation. On August 25, 2002, Hatfill stood in front of his lawyer's
office and told a crowd of news media, "I am not the anthrax
killer . . . I am an innocent man," and the FBI's inability to overcome Hatfill's solid alibi or to link him directly to the crime reinforced Hatfill's claim.[68] Angry that the investigation had cost him

his job and his reputation, Hatfill sued the Justice Department for defamation and eventually won several million dollars.[69]

Pressure mounted on the FBI to find the anthrax killer, but Wilson could not make the genomic analysis move any faster. In a 2003 interview, Harp reminded Americans that fifty FBI investigators were still working full time on the Amerithrax case, backed by "a huge scientific effort" from the associated laboratories. The scientific data just took time to be generated. "We're making progress," he said.[70] The FBI kept looking for the perpetrator, increasingly relying on the slowly emerging genomic evidence about the strain of *B. anthracis* mailed in the deadly letters. Despite tracking every lead, even hoaxes whose powders were sent to USAMRIID for testing, FBI agents found the trail going cold by early 2004.[71]

The End of the Trail

The perpetrator probably had no idea that, in the next few years, the geneticists in Paul Keim's group would develop such sensitive tests that they would be able to pinpoint not only the laboratory, but even the very flask or culture dish from which the bacilli had originated. During the investigation, everyone moved and worked in uncharted territory: the genomic techniques developed rapidly; their fit with the needs of microbial forensics was uncertain; and the complex combination of forensic procedures being used had not been tested in a court of law. No one was sure whether genomic analysis of the attack strain would lead to what Keim called "the forensically meaningful source"—the perpetrator of the crime.[72]

By early 2006, the combined efforts of TIGR and the Keim lab had identified four rare mutations that differentiated the Ames subtype used in the attacks from its closest relatives (also Ames strains). The attack strain now had a unique identity, a unique fingerprint based on these four mutations. If the geneticists could find an identical substrain at USAMRIID or in Keim's library of *B. anthracis* strains, they would know the source of the attack bacilli.[73] Only eight of over one thousand spore samples they tested matched this profile, and one of the eight (what I will call the mother culture) was the source of the other seven. The perpetrator of the letter attacks had taken his spores from this same mother culture, cultivated them separately, gotten them to sporulate again by exposing them to air, and then processed the resulting spores to weaponize them. The discovery of the mother culture, the result of much hard work carried out in secrecy, had tremendous implications for the ability of genomic scientists to re-create the family tree of an organism such as *B. anthracis* as well as for microbial forensics. "We are approaching the theoretical limits for both specificity (single nucleotide) and sensitivity . . . for strain sub-typing," wrote Paul Keim's colleague Matthew Van Ert in early 2007.[74] To the extent that they could without violating their secrecy pledge to the FBI, the geneticists began to submit their evidence to the scrutiny of the larger scientific community. For them, the standard of proof lay with acceptance by their peers, demonstrated by publication in peer-reviewed scientific journals. What Van Ert and Keim did not mention in their 2007 paper was the fact that their analysis had helped the FBI track the attack strain of *B. anthracis* to a single laboratory at USAMRIID, and eventually to a single flask labeled RMR-1029 in one of the secure lab rooms. From this flask, anthrax bacilli had been cultured and processed to make the powders mailed in the deadly letters.

The genomes of the spores in the flask read as a record of human activity: thirteen runs of production at Dugway Proving Ground and twenty-two spore preparations made at Fort Detrick had all left their imprint on the RMR-1029 *B. anthracis.* The mother culture in the RMR-1029 flask contained highly processed and highly concentrated spores that had been through several different culturing procedures. They were a mixture of substrains and could almost be guaranteed to produce many mutants that would create what Paul Keim called "a distinct signature" in any spores that were descendants of the mother culture.[75] (This characteristic made it possible for the Amerithrax investigators to definitively link the attack spores to the mother culture.) The many processes that had shaped the contents of the RMR-1029 flask

gave an identity to the organism that was probably more dependent on past human activity than on the natural attributes of
Bacillus anthracis. What was inside RMR-1029 was very distinctive. The keeper of the flask was a twenty-year veteran of anthrax
vaccine research at USAMRIID, Bruce E. Ivins.[76]

Ivins was well known among anthrax researchers. He had been
on the team that first located the toxin-producing genes on *B. anthracis'* genome. He had been at the leading edge of developing
innovative component vaccines and had even assisted the FBI
with its investigation of the powders found in the deadly letters.
But beginning in 2007, after the bacilli from the envelopes had
been traced to the RMR-1029 flask, Ivins came under intense
scrutiny. He was subjected to constant surveillance, stressful interviews, and search warrants that allowed the FBI to take items
from his home, his laboratory, and his cars.[77] Though FBI agents
also carefully scrutinized the other people who had access to the
RMR-1029 flask, they ruled them out and increasingly focused in
on Ivins. In July 2008, the FBI informed Ivins's lawyer, Paul
Kemp, that his client would soon be indicted on charges related
to the anthrax attacks. On July 29, Ivins swallowed an overdose
of acetaminophen and died, having apparently left behind no
confession or other clear information about the reason for his
suicide.[78]

Ivins's death stunned his colleagues and sent shock waves
through the scientific world, now forcefully confronted by the
likelihood that such a well-known scientific insider had committed the crime. Ivins, a gardener and devoted Red Cross volunteer,
was also a top-notch scientist with a high-level security clearance
who had been a trusted member of the small group that worked
on anthrax. Under very difficult circumstances, Ivins had worked
long hours to assist the FBI investigation and as a result was one
of three USAMRIID scientists awarded the Decoration of Exceptional Civilian Service by the Defense Department in 2003. Yet
the FBI's evidence also revealed Ivins to be suffering from serious
psychological problems. In e-mails and counseling sessions, Ivins

sounded angry and threatening. His extensive private collection of weapons included body armor and a taser. The pressure of his work on Amerithrax and, later, the FBI's intensive scrutiny increasingly got to him, according to friends.[79]

Ivins also had a possible motive for causing a frightening outbreak of premeditated anthrax. He felt passionately that vaccine research and development were critical to national security. Amid the controversy over the mandatory military vaccination program in 1998–2000, Ivins was upset that the production of one of the experimental anthrax vaccines developed at USAMRIID had been cancelled.[80] Gary Mastumoto, the journalist accusing the U.S. military of having given Ivins's experimental vaccine to soldiers, pursued Freedom of Information Act requests for information from Ivins in 2000–2001. This reportedly angered Ivins and, the FBI suggested, may have helped persuade him to mail the anthrax letters.[81] Perhaps Ivins wanted to draw attention forcefully to what he perceived as the critical need to keep producing new anthrax vaccines to safeguard national security. Ivins might not have realized that the spores would disseminate so widely and kill people other than the targets. One former colleague quoted in the *Los Angeles Times,* assuming Ivins was guilty, said, "I don't think he ever intended to kill anybody. He just wanted to prove 'Look, this is possible.' He probably had no clue that it would aerosolize through those envelopes and kill those postal workers."[82] But because Ivins had not left a suicide letter or made any kind of confession, no one could be sure of his motivation.

Some skeptics even doubted that Ivins had been the perpetrator. They pointed out that, with Ivins's death, the FBI's evidence would not be examined in a court of law and could not have its strengths and weaknesses publicly assessed. The FBI's evidence, as presented, was circumstantially very strong but could not definitively implicate Ivins. Ivins had control of the RMR-1029 flask, but others had access to it. Ivins had spent extra hours in the secured lab room for several nights before each of the anthrax mailings had occurred, so he had had access to RMR-1029. But nothing Ivins had said or done indicated any particular reaction to the deaths of Bob Stevens and the other anthrax victims (although he apparently fancied himself a criminal investigator).[83] The FBI could not place him near the mailbox in Princeton, New Jersey, from which the letters were almost certain to have been mailed. Most seriously, FBI agents had not found a single spore of *B. anthracis* in Ivins's cars, house, or possessions, despite seizing and testing numerous swabs taken from those areas. Because of this, the FBI could not complete the chain of causality. Genomics had identified the exact location of the source of the murder weapon—the RMR-1029 flask in Ivins's lab—and the detective work had built up a powerful circumstantial case against him. But to Ivins's supporters, the microbial evidence did not seamlessly connect with the forensic evidence. To them, the missing link between these two aspects of the investigation meant that the FBI had not proved its case.

Was Bruce Ivins's suicide a tacit admission of guilt (as the FBI asserted), or had the pressure of the investigation driven him to his death (as his friends insisted)? Skeptical members of Congress and the scientific community called for an open airing of the FBI's evidence and a detailed description of the genomic science that had found RMR-1029 and implicated Ivins.[84] Four months after his death, Ivins's guilt or innocence were no closer to being resolved. The FBI and Justice Department insisted that the evidence would have been strong enough to convict Ivins, but critics continued to assert that the evidence was circumstantial, and the scientific evidence, although convincing, was untested in court. In response, the FBI asked the National Academy of Sciences to convene a panel of experts to answer questions about the FBI's investigation, which it did. (At this writing, March 2010, the report has not yet been released.) This move did not silence Representative Rush Holt and other critics, however, who continued to advocate a congressional investigation.[85] The Department of Justice officially closed the Amerithrax investigation on February 19, 2010, but the controversy over Ivins' guilt or innocence continued. Barring a new piece of smoking gun evidence or a deathbed confession, we may never be able to resolve definitively the question of who created the fine powder of *B. anthracis* and sent the deadly letters.

The reasons for this failure lie not with conspiracies or mistakes so much as with a disconnect central to the procedures of microbial forensics. This type of investigation differed from the more common uses of scientific evidence in forensics, such as using evidence in the form of the perpetrator's DNA in sexual assault prosecutions. In this case, the murder weapon and the holder of the DNA evidence was a living organism, a microbe. Microbes have their own identities and life cycles and are not materially connected to the perpetrator of a crime. In addition, a microbial forensics investigation is a hybrid of two very different ways of investigating and drawing conclusions. The types of evidence—genomic versus psychological and epidemiological—differ. So do the procedures for validating that evidence: the geneticists relied on scientific substantiation through peer-reviewed publication, whereas the FBI agents depended on making their case in court to prove criminal attribution. Lacking solid missing-link evidence—for example, spores in Ivins's car—the FBI has sought to formulate a verification process that would hold up in court.

Verification—identifying the cause, source, and perpetrator of the premeditated disease—remains incomplete, yet verification is needed to resolve the criminal case. To attain resolution of a case, most or all of the stakeholders needed to agree on verification that someone had deliberately created and used *Bacillus anthracis* as a biological weapon, causing an outbreak of premeditated anthrax. Perhaps the two kinds of evidence did not overlap sufficiently to prove the association between the perpetrator, the bacilli, and the disease outbreak. Or perhaps the stakeholders had not yet found a translational zone that could unite the genomic and criminal procedural ways of knowing such that everyone could be convinced. The deadly letters may very well remain one of the United States' great scientific mysteries—which is perhaps not too surprising given the secrecy in which human–*B. anthracis* interactions have been shrouded for so long.

THE PREMEDITATED DISEASE

The Meselson team's investigation at Sverdlovsk, hoaxes, compulsory vaccinations, and Amerithrax: these events of the 1990s and early 2000s point toward major changes in the identity of anthrax at the beginning of the twenty-first century. In much of the world now, people understand anthrax as the manifestation of premeditated action to create bacilli that are biological weapons. Yet although cases of the so-called natural forms of the disease in agricultural populations, wool-mill workers, and others are rare, the incidence of premeditated anthrax is even rarer. Nonetheless, this identity for anthrax is particularly resonant at this time in history. We need to keep trying to construct a framework of knowing about premeditated anthrax, which may help to prevent future outbreaks.

For historians, the development of microbial forensics has signaled something rather profound. The investigation into the deadly letters mailed to Americans in 2001 demonstrated how the record of change over time in microorganisms is written into the microorganism's very physical makeup. The spores in the RMR-1029 flask taken from USAMRIID embodied how recent human activities had changed *B. anthracis*. This identity was a hallmark that distinguished outbreaks of disease due to natural causes from outbreaks caused by a deliberately seeded pathogenic organism. The level of genomic resolution has now increased to the point to where we can read the history of the human interactions with the organism, and the genome, down to each nucleotide, has become a complex historical text.

In the investigation of the 2001 postal attacks, the science was only one piece of the puzzle, but in the end it provided the most important material evidence for the criminal investigation. This part of the investigation revealed a history of human interactions with the attack strain of the bacillus. It told of the bacilli's travels around the country, from their natural origins in an obscure Texas cow pasture to the halls of Congress in Washington, D.C. The bacilli's brother and sister strains had even traveled the world by way of scientists' practice of swapping cultures with trusted colleagues. The narrative of its life linked Sverdlovsk to the U.S. vaccine program, which was in turn shaped by the climate of fear in the late 1990s and the two Bush administrations' Persian Gulf wars. By mailing a few letters, the perpetrator of the anthrax attacks brought this strain of *B. anthracis* to homes, postal centers, and office buildings around the United States.

The resulting federal investigation helped to create what the *New York Times* called an "unprecedented boom in biodefense research," uniting scientists from military, corporate, and university-based laboratories.[86] The scientists have gained tremendous resources for pursuing the molecular evolution and ecology of *B. anthracis* and other potential biological weapons, but they have had to learn new procedures and even a new culture, that of microbial forensics and criminal investigation, and endure exceptional scrutiny. With forensic investigation of the deadly letters pointing to a member of the close-knit group working with anthrax in the late 1990s, scientists have again confronted moral issues last faced by their predecessors during World War II and the cold war. The contents of the RMR-1029 flask were artifacts, separated from natural context, a deliberately laboratory-created agent that did what its creators had shaped it to do: kill.

The investigations into the 1979 and 2001 anthrax deaths have linked the biology of the new century to that of the old. These links have been embodied in people such as William Patrick, who worked with powdered anthrax during the cold war, and Donald Henderson, the overseer of the World Health Organization's global smallpox eradication program, who encouraged the Meselson team.[87] Procedures for investigating the appearances of the bacillus and the disease also spanned the turn of the century, as both the Sverdlovsk and the FBI investigations during 2001–2010 relied on detective work and epidemiology as well as genetic analysis of the spores.

Incidents of bioterrorism have framed outbreaks of anthrax

in new ways. New constellations of scientists from the FBI, USAMRIID, corporations, and universities are now necessary to query the bacillus and ascertain its provenance. Procedures have developed that now push our ability to identify *B. anthracis* strains down to the molecular level. This framework of procedures, and the scientists who perform them, still coalesces around the venerable and common goal of finding out even more about *B. anthracis* and being able to manipulate it and keep it in check—the need to control anthrax persists. But we must now consider the possibility that an anthrax outbreak is the sad outcome of malevolent human intent and a century of domesticating and manipulating *Bacillus anthracis.* Along with its other identities, anthrax has become a premeditated disease.

Epilogue
Stories about Anthrax

✻ ✻ ✻

While fear remains Camp Detrick and its sister stations throughout the world must go on storing up destruction. If we had peace, these places could show us how to abolish influenza and the common cold, tuberculosis, malaria, and all the other natural plagues of man, as well as those of animals and plants. There is no reason to doubt that these things could be done; but first we must abolish the unnatural plague of war.

Theodor Rosebury, 1949

Diseases change over time through complex and dynamic biological and sociocultural interactions. This book has detailed the process of how humans domesticated *Bacillus anthracis* and the price we continue to pay for having done so. Accidental and deliberate deployments of *B. anthracis* have created new and more deadly anthrax districts—environments not safe for humans or their animals to inhabit. We have invited an organism that is one of our predators into ever more intimate contact with us. Over the past two hundred years, we have expanded *Bacillus anthracis'* habitat and bred it to be more virulent. And we did so at the same time that we sought to control it and its kind—microorganisms—in the name of decreasing their depredations on human populations.

These seemingly opposing goals reflect our varying perceptions and explanations of the purposes that a microorganism can serve. Microorganisms, and human control of their abilities to parasitize and injure us, have been at the center of one of science's great success stories. This narrative is one of progress: greater scientific knowledge and medical control of microorganisms led to less human and animal suffering and death from infectious diseases. During the twentieth century, the development of medical interventions such as serum injections, vaccines, and antibiotics altered the natural course of many acute infectious diseases, shifting the ecological balance toward the survival of humans. This shift has also occurred with anthrax. *B. anthracis* must kill its host to perpetuate its life cycle (very unusual for infectious agents, many of which make the host sick but don't kill it). We, the potential hosts, have defended ourselves by either killing the bacilli (disinfecting imported wool, for example) or seeking to prevent the harm it causes to an individual's body (through vaccines and antibiotics). These social and medical interventions have largely rendered the old, natural forms of anthrax—the agricultural and industrial outbreaks of cutaneous and gastrointestinal disease—rare and, from a public health perspective, insignificant in the developed world.

But even as we have accomplished this defense, we have used microorganisms as tools of destruction. Doing so has created a new identity for some diseases that I have called premeditated; outbreaks of these diseases could result from weaponized microorganisms. Outbreaks of premeditated anthrax are thus markers of the presence of weaponized *B. anthracis,* and its story is one of the stories that this book tells about anthrax. It is not a story of progress, however, in the sense of controlling infectious disease. Rather, the same methods scientists had used in working with the bacillus for "good purposes" could also be used to create new deadly manifestations of the disease. During the late 1800s, *B. anthracis* was domesticated in the laboratory, and factory workers died of woolsorters' disease (inhalational anthrax). These historical events formed the basis of scientists' focus on *B. anthracis* for weaponization in the early to mid-1900s.

In the twentieth century, scientists saw that the very qualities of the bacillus that they were researching for the sake of improving human health were qualities that made it weaponizable. In 1949, American microbiologist Theodor Rosebury asked, "What distinguishes a potential [biological weapons] agent from just any germ?" He cited "infectivity; casualty effectiveness; availability; resistance; means of transmission; specific immunization; therapy; detection; and retroactivity" as the major characteristics that would make a microorganism suitable to become a biological weapon. *Bacillus anthracis* had most of them.[1] But these characteristics were only the starting point for the process of making some strains of *B. anthracis* into weapons.

The standard story about weaponized *B. anthracis* tells us that the power of modern science has utterly altered the bacillus and made it unnaturally virulent and infective. What scientists have long referred to as wild-type *B. anthracis* came into the laboratory in samples of animals' and humans' bodily fluids and parts. Many philosophers, Karl Marx included, would have described the bacilli contained in this material as first nature, something that came from nature and was not human made. After living in laboratories, being cultivated, treated, and passaged through the bodies of experimental animals, however, the bacillus became something new: second nature, a product of human activity.[2] It could no longer be considered the same thing as the bacilli swarming in the carcasses of dead animals out in the field; it was a creature of artifice and not nature. Certainly the Ames, Sterne, and Vollum strains bred by scientists fit this definition. Those strains could then be further manipulated by scientists to be cured of their virulence by removing particular genes in order to make vaccines against them. These cured strains could no longer cause anthrax; they were mere tools for human use. This narrative of creating artificial *B. anthracis* thus illustrates the power of scientists' knowledge and practices to manipulate microorganisms and make them behave in particular ways.

It is also a narrative that the stories in this book complicate. First of all, so-called artificial strains of *Bacillus anthracis* retained many of their material, or natural, characteristics. These characteristics formed the boundaries within which scientists worked when they brought *B. anthracis* into the laboratory. For example, *B. anthracis'* basic life cycle did not change; nor did many of the nutritional requirements peculiar to this organism. Second, the laboratory bacilli often acted in ways that their "creators" would not have predicted. They sometimes lost or gained their killing powers at times unexpected by their human handlers. They infected or killed laboratory workers, and sometimes they escaped to kill animals and people outside the laboratory. In short, the bacilli behaved in ways not controlled by their human handlers: they were not controllable. All of this points to *B. anthracis*, in its many manifestations, as having some qualities that were human made and other qualities that were not. A hybrid creature, it nonetheless has retained original characteristics of its identity both within and outside human control. Some of these characteristics have been preserved due to their usefulness to its human manipulators; but others have persisted despite us.

These hybrid creatures could be problematic in the long term. Their creators could no more easily return them to their natural state than they could put the proverbial genie back in the bottle. The relationships between human and microbial populations had been forever changed, and some scientists feared the consequences. "It cannot be too urgently stated that the advance of science is allowing the development of wholly unnatural modes of human domination and mass destruction," warned David O. White and Macfarlane Burnet in the early 1970s. "Unnatural in the sense that they represent conditions never encountered in the course of human evolution. . . [No] mammalian species [has had] to face the impact of a pathogenic microorganism that had not evolved . . . by natural processes."[3] Burnet, a Nobel Prize–winning physician and virologist, believed that the consequences of "artificial evolution" could threaten "man's relationship to his global ecosystem," for example, upsetting the balanced relationship between *B. anthracis* and its hosts that had developed over thousands of years.[4] In more recent years, another Nobel laureate, Joshua Lederberg, has echoed Burnet's concerns. These scientists and others remain troubled that twentieth-century political and military decisions made some scientists into weapons designers and some strains of *B. anthracis* and other microorganisms into weapons of mass destruction.

ETHICAL DILEMMAS

Between World War I and the 1980s, research programs dedicated to finding defensive measures against *B. anthracis* also often produced knowledge about the microorganism that helped make it into a weapon. As one journalist put it in 1995, "The perfectly legitimate endeavor of making anthrax vaccine . . . is an excuse for having anthrax around—one of several 'masks' for weapons production."[5] The dual nature of *B. anthracis* research—and its role as a link between American bacteriology and the American military—provided a space of negotiation for several conflicts: the appropriate role for science in military affairs, the morality of using technology and ideas designed to save lives for the taking of lives, the police power of governments to involve ever larger numbers of citizens (for example, forcing soldiers to be vaccinated against *B. anthracis*). In addition, the availability and affordability of biological agents have made them attractive to an ever-widening circle of rogue states, groups, and individuals bent on terrorism. This reality has contributed to the expansion of the science-military relationship to include larger numbers of non-military bacteriologists and to rely on their expertise and their willingness to work in secrecy. Asked to use his Northern Arizona State laboratory to advance the forensic investigation of the anthrax letters in 2001, geneticist Paul Keim acknowledged that "everything's changed."[6]

Change came with a price. People in places like Sverdlovsk, and the countless animals sacrificed to testing with *B. anthracis,* paid the greatest price. Other peoples' environments became contaminated or were used to calibrate doses of *B. globigii* without the inhabitants' consent or knowledge. Working with biological weapons forced scientists to think "upside down" and arguably altered scientific culture in the twentieth century. New boundaries between traditional tenets of scientific culture had to be negotiated when scientific work was done within the constraints of military oversight and in the name of national security. Openness, cooperation, and sharing of biological material, crucial to the enterprise of science, were often prohibited in the context of biological weapons work.

Bacteriologists, pathologists, physicians, and veterinarians struggled with the fact that biological weapons research required them to work toward goals of destruction rather than the preservation of human and animal life. For many, biological weapons were more morally repugnant than nuclear weapons. Biological weapons attacked from within; they were foreign to traditional modes of warfare and codes of honor that allowed someone to know against whom he or she was fighting. Creating modes of death directly violated physicians' and veterinarians' honor codes and the oath they all took to "do no harm." Nonmedical scientists felt uneasy too. Theodor Rosebury summarized the situation this way: "How can science aimed at the destruction of human life still cherish the notion that human life is the highest value?"[7] From scientists' and physicians' calling to preserve human lives had come the terrible ability to use scientific and medical skills to design mass suffering and death. Sociologist Jeanne Guillemin, reflecting on the deaths in Sverdlovsk, wrote that "out of biology, the science of life and healing, have come weapons that can turn the worst of nature's devices against humans. . . . Biological weapons . . . represent a hideous assault on life."[8]

Why were scientists willing to work on creating weapons? One story told by scientists was that the Allies during World War II (Britain, France, the United States, and other nations) feared that the enemy Axis powers (Germany, Japan, and others) would create a biological weapon first and deploy it against Allied soldiers or civilians. This was not an unreasonable belief, given the rumors about events in Manchuria (Manchuko) and the known history of German agents using anthrax against animals during World War I. Even the most ethically concerned scientists who worked on weaponizing *B. anthracis* during World War II (including Rosebury, a dedicated pacifist) put their reservations aside and responded to the desperate circumstances of war and fear of "imminent danger" from biological weapons created by an enemy.[9] With historical hindsight, we now know that the threat of Axis weapons use was never realized. Japanese formulations of *B. anthracis* and its deployment strategies were crude and (as far as we know) used mainly within Manchuria; German scientists did not get very far at all in developing weaponized *B. anthracis.* Yet the narrative of imagined enemy production of *B. anthracis* drove Allied scientists to try to transform this microorganism into a biological weapon during wartime. Deterrence and defense got couched in the terms of parity in biological weapons research and production.

If the war was such an important justification, then why did these programs continue after the war ended? Many scientists left the weapons research and production facilities after the war, but others stayed. British microbiologist David Henderson, who continued to work at the MDE at Porton Down during the cold war, saw his work as a natural outgrowth of the wartime program. According to Henderson, scientists had been under too much time pressure during the war; now they had the leisure to do the experiments right and to continue the creation of basic knowledge about inhalation biology. The apparatus developed during the war (the piccolo) could now serve its purpose for *science,* and the results of the work could be openly published and discussed with other scientists. As this story went, experimental work could still be interesting while focusing on defensive, rather than offensive, formulations of the biological agents. Henderson and the others were able to participate in the scientific community while still working in the research areas they had developed during the war. Finally, the framework of the cold war reinforced the need for continued research on *B. anthracis* and other potential weapons agents, and the secrecy surrounding these projects helped to ensure their continuation in the form of basic research even after offensive weapons development had supposedly ended in the early 1970s.

One model that may help us to understand why Henderson and other scientists continued working at the junction of basic research and national security is a shadowy American group of scientists called the Jasons. The Jasons founded themselves around 1960 as a group of (mainly theoretical) physicists interested in advising the Department of Defense (DoD) on various problems of national security. The group would meet for six weeks every summer and try to come up with creative solutions for the problems of how to shoot down missiles or improve submarine radar technology. Since then, the group has broadened its perspectives to include climate change and basic biological modeling issues; environmental scientists and biologists, at least one of whom has written about biological warfare, have joined the core of physicists.

Similar to the sentiments of cold war bioweapons scientists such as David Henderson and William Patrick (a member of the U.S. scientific program), the Jasons relished working on applied projects, "more like an engineer," as one Jasonite put it. The projects seemed "sensible and important," a chance to take a hands-on approach using the scientists' theoretical abilities.[10] The solutions occasionally included designing an actual apparatus—much like Henderson's group, who seemed to enjoy tinkering incessantly with their piccolo. Social benefits followed, too: scientists remained close personally and professionally even through heated disagreements (and, in the case of the *B. anthracis* researchers, even after the end of World War II). They enjoyed working with smart and creative people and working as a group. The most fascinating parallel, however, is the persistence of patriotism as a motivation for applying science to national priorities. "Being a patriot" is perhaps the most meaningful story for scientists who have worked in weapons programs (either defensive or offensive). As one Jasonite explained, "I am a scientist. But before I am a scientist, I'm a U.S. citizen . . . [if] I have a problem with what the DoD is doing, then it is my responsibility to protest that . . . [but] I think that DoD is an important part of our society." She said she suffered no "challenges to my conscience" for being the recipient of research money from the DoD or for advising it as a Jasonite.[11] Even scientists who thought carefully before joining felt that "there was always the possibility of doing some good" for the nation.[12]

Like the Jasonites, many biological scientists who continued to work on *B. anthracis* (including the highly virulent strains originally created to be weapons) felt that they were doing patriotic work. Colleagues remembered the scientist Bruce E. Ivins, accused of mailing the American anthrax letters in 2001, as a patriot concerned about the safety of American soldiers and civilians who might be exposed to weaponized *B. anthracis.* One story about a possible motivation for Ivins to have sent the letters cast him as wanting to draw attention to decreasing government funding for anti-anthrax vaccine research. If ordinary Americans feared contracting premeditated anthrax through the mail, then pressure would be put on agencies such as the National Institutes of Health and the National Science Foundation to restore higher levels of funding for *B. anthracis* research (which happened to be Ivins's area of expertise). If the perpetrator of the letter attacks sought to increase consumer demand for anti-anthrax vaccines, the story goes, he could well have done so under the guise of patriotism. Sending the anthrax letters would serve both purposes.[13]

Was *B. anthracis* a credible threat to the collective security of North Americans (and, by implication, to the security of other people around the world)? There are good arguments for both "yes" and "no" responses. In a very real sense, however, what matters is the *idea* that *B. anthracis* is dangerous and anyone could be injured by it. Here is where the life cycle of *B. anthracis* intersects with the capacity to create fear. What could be more frightening than an organism that must kill its mammalian host to propagate itself? And one that stays around in the soil for years, continually capable of infecting and killing fresh hosts? The fear and anxiety associated with premeditated anthrax have created far greater effects than the depredations of the bacillus. As one saying went in 2001, "Anthrax isn't contagious, but fear is."[14] The narrative of fear mandates that we respond to forces over which we may have little control, whether we encounter *B. anthracis* by herding livestock or by opening the mail.

In 2001, Americans responded immediately when spore-laced letters began appearing in media outlets' mailrooms and in congressional offices on Capitol Hill. Politicians and government employees in Washington, along with journalists in newsrooms up and down the East Coast, fought what one senior Bush administration official called "a war of nerves" fraught with "confusion and anxiety, verging on terror" as hundreds of people lined up to be tested for exposure to *B. anthracis*.[15] Ordinary people around the country also struggled to maintain a sense of perspective. U.S. Senator Joe Biden was surprised when his constituents wanted only to ask him about what he was doing "to protect my kids against terrorists." He summarized the transformed priorities of American women this way: "Soccer moms are security moms now." Half of American women and one-third of American men polled nationally in October 2001 believed that a member of their family would soon be a victim of some kind of terrorist attack.[16] People with absolutely no reason to worry about exposure to the spore-bearing letters demanded ciprofloxacin from their physicians (also called cipro, a prescription drug known to be effective against *B. anthracis*). An article in the popular *Time Magazine* exhorted Americans to "Be a Patriot. Don't Hoard Cipro!"[17] Other articles in the popular press gave instructions on how to manage "anthrax anxiety," remain calm, and continue with daily life.[18] But clearly, American citizens demanding cipro imagined themselves having already having been exposed—*have I opened any letters that contained powder or smelled strange?*—and perhaps suffering terrible illness and death.[19] This intensity of fear waned as the months passed, of course, but the anthrax letters attack brought ideas about risk into sharp focus in 2001.

HISTORY AND THE NARRATIVE OF RISK

Premeditated anthrax is, in some senses, an imagined epidemic. We imagine it even when it is not present. Although the risk of contracting premeditated anthrax is small, the appearance of the disease is dramatic. Outbreaks of premeditated anthrax have only rarely appeared in human populations. Very few have died from it compared with other infectious diseases (such as HIV/AIDS or malaria). Yet some nations (such as the United States) spend a great deal of money and effort every year preparing for outbreaks of premeditated anthrax, researching new types of vaccines and monitoring the appearance of suspected cases. Why, when cases of anthrax are so rare?

History points to two important explanations. First, humans' past experiences with anthrax have created a powerful story of *B. anthracis* (and the disease it causes) as a deadly and ruthless killer, lying in wait in cursed and infected environments for unsuspecting human victims. Second, by coming up with a story about why we *should* fear anthrax—a narrative of risk—we try to make sense of our world and our vulnerability to human malice. The particular horror of a premeditated outbreak from a weaponized form of *B. anthracis* lies in the fact that a fellow human wished us harm and used a microorganism to carry his deadly message. The outcome of policies to create and deploy *B. anthracis* is based on the assumption that our enemies at the time are dangerously close to controlling a biological weapon. In the past, the narrative of risk from the Western point of view focused on the Axis powers and then on the Soviet Union; now, rogue regimes and shadowy terrorist groups are the enemies. It remains to be seen if premeditated anthrax will be a favored tactic for terrorists in the future.

Historians do not possess crystal balls, and we are no more likely to be able to predict the future than anyone else. But one thing seems clear: the bacillus' biological characteristics and the associated narrative of risk have sealed the identity of anthrax as a premeditated disease. Many future outbreaks will be suspected of being premeditated. *B. anthracis* will probably remain useful as a biological weapon in part because it is unlikely to evolve toward a more benign relationship with populations of susceptible people (who are equally unlikely to evolve ways to evade *both* of its lethal toxins in the near future) and in part because narratives of fear and risk are remarkably persistent.

In the end, we owe it to the victims of Sverdlovsk, Manchuko, Zimbabwe, the 2001 U.S. postal attacks, and other presumed premeditated outbreaks to examine carefully our continuing relationship with biological weapons through the lenses of ethics and history. We have domesticated and shaped *Bacillus anthracis,* expecting that we could control it. But the bacillus must kill to survive, and any time we welcome it into our *domus,* we must expect that it will continue to act accordingly. And now we come again to what physician René Dubos said in 1959: "Man himself, like the sorcerer's apprentice, has set in motion forces that are potentially destructive and may someday escape his control."[20] In the case of anthrax, that day has come.

ACKNOWLEDGMENTS

A book is a collaborative effort, and I could not have written this one without the advice of many people. I am especially indebted to Charles Rosenberg, Jacqueline Wehmueller, and an anonymous reviewer (all of whom read and improved the manuscript), and Philip M. Teigen and Martin Hugh-Jones for their ideas and encouragement. For stimulating conversations, I thank my colleagues from the Program in the History of Science, Technology and Medicine; the Institute for Advanced Study at the University of Minnesota; and commentators and questioners from audiences around the world. The timely completion of this project depended on Karin Matchett's editing and the research assistance of Christine Manganaro, Cara Kinzelman, and Kelly Lusk.

My research has been supported by the Universities of Minnesota and Colorado, and by a grant from the U.S. Department of Homeland Security's National Consortium for the Study of Terrorism and Responses to Terrorism. I also thank archivists and staff, who directed me to important sources of evidence in the following libraries. In the United Kingdom: Bradford Central Library Archives and Local History Collections; Bradford University Libraries; the Wellcome Library (History of Medicine), London; the Public Records Office (Kew); and the British Library, London. In the United States: the National Archives and Records Administration (College Park, Suitland); the Library of Congress; the National Library of Medicine (History of Medicine Division); and the Wangensteen Library for the History of Medicine and the College of Veterinary Medicine Library at the University of Minnesota.

Finally, words are inadequate to express my thanks to my husband, Kevin R. Reitz, and son, Phil Reitz-Jones, who have cheerfully reminded me of what's really important in life.

NOTES

Introduction

1. My thanks to Jennifer Gunn for bringing this point to my attention.

2. Laurie A. Wilkie, *The Archaeology of Mothering: an African-American Midwife's Tale* (New York: Routledge, 2003), xxi.

3. Matthew N. Van Ert, W. Ryan Easterday, Lynn Y. Huynh, Richard T. Okinaka, Martin Hugh-Jones, Jacques Ravel, Shaylan R. Zanecki, Talima Pearson, Tatum S. Simonson, Jana M. U'Ren, Sergey M. Kachur, Rebecca R. Leadem-Dougherty, Shane D. Rhoton, Guenevier Zinser, Jason Farlow, Pamela R. Coker, Kimothy L. Smith, Bingxiang Wang, Leo J. Kenefic, Claire M. Fraser-Liggett, David M. Wagner, and Paul Keim. "Global Genetic Population Structure of *Bacillus anthracis*," *PLoS One* 2, 5 (May 2007): e461, 1–10.

4. Robert E. Kohler, *Lords of the Fly:* Drosophila *Genetics and the Experimental Life* (Chicago: University of Chicago Press, 1994); Karen Rader, *Making Mice: Standardizing Animals for American Biomedical Research, 1900–1955* (Princeton, NJ: Princeton University Press, 2004).

5. William H. McNeill, *Plagues and Peoples* (New York: Doubleday, 1976), esp. 224.

6. Bruno Latour, *The Pasteurization of France,* trans. Alan Sheridan and John Law (Cambridge, MA: Harvard University Press, 1988).

7. K. Codell Carter, "The Koch-Pasteur Dispute on Establishing the Cause of Anthrax," *Bulletin of the History of Medicine* 62 (1988): 42–57.

8. Brian Balmer, "How Does an Accident Become an Experiment? Secret Science and the Exposure of the Public to Biological Warfare Agents," *Science as Culture* 13, 2 (June 2004): 197–228.

Chapter 1. Infectivity and Fear

Epigraph: Louis Leonpacher, diary 1908–1923, pp. 845–46, mss. 2141, Special Collections, Louisiana State University Libraries, Baton Rouge. My thanks to Phyllis Hickney Larson and Robert J. Leonpacher for drawing my attention to this diary.

1. Nsekuye Bizimana, *Traditional Veterinary Practice in Africa* (Eschborn: Deutsche Gesellschaft für technische Zusammenarbeit, 1994), 420–21, 487–88; George Fleming, *Animal Plagues: Their History, Nature and Prevention* (London: Chapman and Hall, 1871), 524. I thank Akke Katsma for the Friesian translation.

2. Jean Blancou, *History of the Surveillance and Control of Transmissible Animal Diseases*, trans. from French edition, *Histoire de la Surveillance et du Contrôle des Maladies Animales*, 2000 (Paris: Office International des Épizooties, 2003), 81.

3. My thanks to Professor Peter Koolmees for this information.

4. John Spear, "On the So-Called 'Woolsorters' Disease' as Observed at Bradford and in Neighbouring Districts in the West Riding of Yorkshire," appendix A, no. 8, *Tenth Annual Report of the Local Government Board, 1880–81: Supplement containing the Report of the Medical Officer for 1880* (London: George E. Eyre and William Spottiswoode, for Her Majesty's Stationery Office, 1881), 66–135. See 113, 134–35, quotations on 135.

5. C. D. Stein, "Anthrax in Livestock in the United States and Its Control," *Veterinary Medicine* 48 (Jan. 1953): 8–9.

6. Lise Wilkinson, *Animals and Disease: An Introduction to the History of Comparative Medicine* (Cambridge: Cambridge University Press, 1992), 124–25.

7. David M. Morens, "Characterizing a 'New' Disease: Epizootic and Epidemic Anthrax, 1769–1780," *American Journal of Public Health* 93, 6 (June 2003): 886–93, see 888–89.

8. [Jean, fl. or Nicolas] Fournier, *Observations et Expériences sur le Charbon Malin: avec une Méthode Assurée de la Guérir* (Dijon, France: Defay, 1769), esp. 43–62 (my translation).

9. Richard Swiderski, *Anthrax: A History* (Jefferson, NC: McFarland, 2004), 11.

10. Wilkinson, *Animals and Disease,* 126.

11. Philibert Chabert, *Description et Traitement du Charbon dans les Animaux* (Paris: Imprint Royale, 1780); François Vallart, "Les Épizootie en France de 1700 à 1850," *Histoire et Sociétes Rurales* 14 (2000): 67–104, esp. 91. The clostridial disease, known in English as "blackleg," remained classified as a form of anthrax until the identification of the separate organisms in the 1870s and '80s.

12. Fleming, *Animal Plagues,* 76.

13. Toon P. Wijgergangs, "Tongblaar in de vorige eeuwen," in *Over beesten en boeken. Opstellen over de geschiedenis van de diergeneeskunde en de boekwetenschap,* ed. Koert van der Horst, Peter A. Koolmees, and Adriaan Monna (Rotterdam: Erasmus Publishing, 1995), 45–58. My thanks to Peter Koolmees for bringing this source to my attention and translating it into English.

14. Blancou, *History of the Surveillance,* 79 and 82.

15. Vallart, "Les Épizootie en France," 87.

16. Frank Somers, "Anthrax Amongst Wild Animals in Captivity," *Veterinary Record* no. 1181 (Feb. 25, 1911): 543–44.

17. Blancou, *History of the Surveillance,* 81. The author states, "for a long time, the zoonotic nature of the disease . . . and the fact that it is telluric . . . went unrecognized" (79). This is the case with Western physicians and natural philosophers, but, I argue, not for most observers. Ideas of anthrax-like diseases as a zoonosis, based in certain land areas, have appeared in the literature for the past three millennia.

18. Talima Pearson, Joseph D. Busch, Jacques Ravel, Timothy Read, Shane D. Rhoton, Jana M. U'Ren, Tatum S. Simonson, Sergey M. Kachur, Rebecca R. Leadem, Michelle L. Cardon, Matthew N. Van Ert, Lynn Y. Huynh, Claire M. Fraser, and Paul Keim. "Phylogenetic Discovery Bias in *Bacillus anthracis* using Single-Nucleotide Polymorphisms from Whole-Genome Sequencing," *Proceedings of the National Academy of Sciences* 101 (Sept. 14, 2004): 13536–41.

19. The Bible passage appears in the book of Exodus, chapters 7 and 8. On interpretations by scientists and historians, see J. F. Smithcors, *Evolution of the Veterinary Art: A Narrative Account to 1850* (Kansas City, MO: Veterinary Medicine Publishing, 1957), 30; Doris M. Klemm and William R. Klemm, "A History of Anthrax," *Journal of the American Veterinary Medical Association* 135 (Nov. 1, 1959): 458–62, see 458; Robert Dunlop and David Williams, *Veterinary Medicine: An Illustrated History* (St Louis, MO: Mosby, 1996), 68.

20. A. Krishnaswany, "What a Veterinarian Can Learn from a Study of the Ancient Indian Veterinary Lore," *Indian Veterinary Journal* 21(1945): 388–90; S. Datta, "The Nature and Significance of Veterinary Problems in Ancient and Modern India," *Indian Veterinary Journal* 25 (1948): 115–20. Smithcors points out that Indian authorities of the time had a far broader and more holistic vision of animal healing than did the Greeks, who restricted their animal healing systems to certain diseases and injuries of horses (Smithcors, *Evolution of the Veterinary Art*, 31).

21. J. R. C. Martyn, "Juvenal 2.78-81 and Virgil's Plague," *Classical Philology* 65, 1 (Jan. 1970): 49–50, see 49.

22. Raffaele Roncalli Amici, "Virgil—The Georgics: Veterinary Reflections" (Virgilio—Le Georgiche: Riflessioni Veterinarie), Keynote address, 35th International Congress of the World Association for the History of Veterinary Medicine, Torino, Italy, Sept. 8, 2004.

23. A. Montovani, S. Prosperi, A. Seimenis, and D. Tabbaa, "Mediterranean and Zoonoses: A Relationship" (Mediterraneo e Zoonosi: una Relazione), 35th International Congress of the World Association for the History of Veterinary Medicine, Torino, Italy, Sept. 8, 2004. The authors hailed from Greece, Italy, and Syria.

24. Dunlop and Williams, *Veterinary Medicine*, 181.

25. Ibid., 183.

26. Fleming, *Animal Plagues*, 33.

27. Blancou, *History of the Surveillance*, 80.

28. Fleming, *Animal Plagues*, 74, 94.

29. Klemm and Klemm, "A History of Anthrax," 459

30. Fleming, *Animal Plagues*, 190, 192.

31. F. H. Gilbert, *Recherches sur les Causes des Maladies Charbonneuses dans les Animaux; leurs Caracteres, les Moyens de les Combattre et de Prevenir* (Paris: l'Imprimerie de Republique, 1795). This was a short pamphlet, often published along with other *recherches* on topics such as artificial prairie and the propagation of Merino sheep.

32. P. Keim, L. B. Price, and A. M. Klevytska, "Multiple-Locus Variable-Number Tandem Repeat Analysis Reveals Genetic Relationships within *Bacillus anthracis*," *Journal of Bacteriology* 182 (May 2000): 2928–36.

33. Matthew N. Van Ert, W. Ryan Easterday, Lynn Y. Huynh, et al., "Global Genetic Population Structure of *Bacillus anthracis*," *PLoS One* 2, 5 (May 2007): e461, 1–10, see 7.

34. N. F. Williams, "Anthrax," *Journal of the American Veterinary Medical Association* 81, 1 (1932): 9–25.

35. For more on the early natural and political history of Saint Domingue, see James E. McClellan III, *Colonialism and Science: Saint Domingue in the Old Regime* (Baltimore: Johns Hopkins University Press, 1992).

36. M-P Justin, *Histoire Politique at Statistique de l'Île d'Hayti, Saint-Domingue, Ècrite sur les Documents Officials et des Notes Communiqués par Sir James Barskatt, Livre Troisième* (Paris: Brière, 1826), 109–25. This source is translated and extensively quoted in David M. Morens, "Epidemic Anthrax in the Eighteenth Century, the Americas," *Emerging Infectious Diseases* 8 (Oct. 2002): 1160–63.

37. Ralph Davis, *The Rise of the Atlantic Economies* (London: Weidenfeld and Nicholson, 1975), 45.

38. Morens, "Epidemic Anthrax," 1161.

39. Swiderski, *Anthrax,* 81; McClellan, *Colonialism and Science.*

40. Cited by Morens in "Epidemic Anthrax," these sources include Henri Bertin, "Relation de Quelques Accidens Extraordinaries, Observés à la Guadeloupe, sur les Nègres du Quartier de la Capesterre à la suite de l'usage qu'ils ont fait de la chair des Animaux Morts d'une Maladie épizootique," in É.-M. de Montigny, *Instructions et Avis aux Habitans des Provinces Méridionales de la France, sur la Maladie Putride et Pestilentielle qui Détruit le Bétail* (Paris: Imprimerie Royale, 1775), 49–70; and C. Arthaud, "Épizootie de Saint-Domingue en 1774 and 1775," *Recherches, Mémoires et Observations sur les Maladies Épizootiques de Saint-Domingue,* ed. Arthaud, Cercle des Philadelphes (Cap-François, Saint-Domingue: L'imprimerie Royale, 1788), 11–18.

41. Henri Bertin, "Relation de Quelques Accidens Extraordinaries."

42. Fleming quotes "the learned historiographer of Barbadoes, Reverend Mr Hughes." This is probably Griffith Hughes, author of *The Natural History of Barbados* (published for the author in London in 1750); this text is from Fleming, *Animal Plagues,* 522.

43. M. Worlock, "Memoire sur la Maladie Epizootique Pestilenetielle de l'ile Saint-Domingue," *Recherches, Mémoires et Observations sur les Maladies Épizootiques de Saint-Domingue,* ed. Arthaud, Cercle des Philadelphes (Cap-François, Saint-Domingue: L'imprimerie Royale, 1788), 162–70.

44. Van Ert et al., "Global Genetic Population Structure," 7.

45. J. Kercheval, "On a Remarkable Disease among Cattle and its Propagation to the Human Species," *American Farmer* 45 (1824): 357. This article is also quoted

in full in Robert P. Hanson, "The Earliest Account of Anthrax in Man and Animals in North America," *Journal of the American Veterinary Medical Association* 135 (Nov. 1, 1959): 463–65.

46. Kercheval, "On a Remarkable Disease among Cattle," 357.

47. James Mease, "On Diseases and Accidents of Farmers," *Memoirs of the Philadelphia Society for Promoting Agriculture,* vol. 5 (Philadelphia: Benjamin Warner, 1826), 251.

48. Fleming, *Animal Plagues,* 391–92.

49. Ibid., 536–37. Fleming cites [Bernhard] Laubender, *Uber den Milzbrand der Hausthiere und seine Geschichte* (pamphlet published in Munich, 1814–1815), 163.

50. O. Bollinger, "Infectionen durch thierische Gifte: Milzbrand," *Handbuch der Chronischen Infectionskrankheiten* (Leipzig: Verlag von F.C. W. Vogel, 1874); Bollinger, "Anthrax," in H. von Ziemssen, A. H. Buck, and G. L. Peabody, *Cyclopedia of the Practice of Medicine* (New York: William Wood, 1874–1881), vol. 3 (1875), 372–430, see 379.

51. Louis Pasteur, "On the Etiology of Anthrax," *Comptes Rendus de l'Académie des Sciences* 91 (1880): 86–94, quotation on 87.

52. Glenn Van Ness and C. D. Stein, "Soils of the United States Favorable for Anthrax," *Journal of the American Veterinary Medical Association* 128 (Jan. 1, 1956): 7–9; Jason K. Blackburn, Kristina M. McNyset, Andrew Curtis, and Martin E. Hugh-Jones, "Modeling the Geographic Distribution of *Bacillus anthracis,* the Causative Agent of Anthrax Disease, for the Contiguous United States using Predictive Ecologic Niche Modeling," *American Journal of Tropical Medicine and Hygiene* 77, 6 (2007): 1103–10.

53. For more on Pettenkofer's role in public health, see George Rosen, *A History of Public Health,* expanded ed. (Baltimore: Johns Hopkins University Press, 1993), 155, 234–35. See also Max von Pettenkofer, "Boden und Grundwasser in ihren Beziehungen zu Cholera und Typhus," *Zeitschrift für Biologie* 5, 2 (1869): 171–310.

54. Michael Worboys, *Spreading Germs: Disease Theories and Medical Practice in Britain, 1865–1900* (Cambridge: Cambridge University Press, 2000), 38–39.

55. Soil has been explored as a factor in the development of other diseases as well, especially chronic conditions such as osteomalacia and cardiovascular disease. See, for example, several articles in *Cellular and Molecular Life Sciences* 43 (Jan. 1987).

56. See especially E. D. Weinberg, "The Influence of Soil on Infectious Disease," *Cellular and Molecular Life Sciences* 43 (Jan. 1987): 81–87; Glenn B. Van Ness, "Ecology of Anthrax," *Science* 172 (June 25, 1971): 1303–1307; Van Ness, S. A. Plotkin, R. H. Huffaker, and W. G. Evans, "The Oklahoma-Kansas Anthrax Epizootic of 1957," *Journal of the American Veterinary Medical Association* 134 (Feb. 1, 1959): 125–29; Van Ness and Stein, "Soils of the United States Favorable for Anthrax."

57. Glenn B. Van Ness, "Soil Relationship in the Oklahoma-Kansas Anthrax Outbreak of 1957," *Soil and Water Conservation* 1 (1959): 70–71, see 70. The term *geophytic* generally refers to plants that bud underground.

58. G. Schmid and A. Kaufmann, "Anthrax in Europe: Its Epidemiology, Clinical Characteristics, and Role in Bioterrorism," *Clinical Microbiology and Infectious Diseases* 8 (2002): 479–88, see 480.

59. Van Ness, "Ecology of Anthrax," 1306.

60. Ibid., 1304.

61. Van Ness et al., "The Oklahoma-Kansas Anthrax Epizootic of 1957."

62. Michèle Mock and Tâm Mignot, "Anthrax Toxins and the Host: A Story of Intimacy," *Cellular Microbiology* 5, 1 (2003): 15–23.

63. K. L. Smith, V. De Vos, H. B. Bryden, M. E. Hugh-Jones, A. Klevytska, L. B. Price, P. Keim, and D. T. Scholl. "Meso-scale Ecology of Anthrax in Southern Africa: A Pilot Study of Diversity and Clustering," *Journal of Applied Microbiology* 87 (1997): 204–207, see 206.

64. Charles E. Rosenberg, "Introduction," *Framing Disease: Cultural Studies in Cultural History,* ed. Rosenberg and Janet Golden (New Brunswick, NJ: Rutgers University Press, 1992), xiii.

65. Morens, "Characterizing a 'New' Disease," 888–89.

Chapter 2. Availability

Epigraphs: Bruno Latour, *The Pasteurization of France,* trans. Alan Sheridan and John Law (Cambridge, MA: Harvard University Press, 1988), 80; Henri Bouley, "La nouvelle vaccination," *Revue Scientifique* 29, 10 (1881): 546–50, quotation on 547 (also quoted in Latour, *Pasteurization of France,* 74).

1. Peter C. B. Turnbull, "Introduction: Anthrax History, Disease and Ecology," in *Anthrax,* ed. T. M. Koehler (Berlin: Springer-Verlag, 2002), 1–19, see 2; Ferdinand Cohn, "Untersuchungen über Bacterien. II," *Beiträge zur Biologie der Pflanzen* 1, 3 (1875): 141–207.

2. K. Codell Carter, *The Rise of Causal Concepts of Disease: Case Histories* (Burlington, VT: Ashgate, 2003), 62.

3. Michael Worboys, *Spreading Germs: Disease Theories and Medical Practice in Britain, 1865–1900* (Cambridge: Cambridge University Press, 2000), 194.

4. Latour, *The Pasteurization of France,* 74–80.

5. Angela N. H. Creager, *The Life of a Virus: Tobacco Mosaic Virus as an Experimental Model, 1930–1965* (Chicago: University of Chicago Press, 2001), 48; Robert E. Kohler, *Lords of the Fly:* Drosophila *Genetics and the Experimental Life* (Chicago, IL: University of Chicago Press, 1994).

6. For an in-depth discussion of causation theories, see Carter, *The Rise of Causal Concepts of Disease.*

7. David Morens likewise describes a "process" of understanding anthrax, but he describes the theorizing tradition as "scientific progress," an erroneous characterization in my view. See "Characterizing a 'New' Disease: Epizootic and Epidemic Anthrax, 1769–1780," *American Journal of Public Health* 93 (June 2003): 886–93, see 886.

8. Aristotle, *History of Animals,* book 4, chapter 8; Jean-Jacques Paulet, *Recher-*

ches historiques at physiques sur les maladies épizootiques avec les moyens d'y remédier,
dans tous les cas, 2nd ed. (Paris: Ruault, 1775), 331, cited in Jean Blancou, *History of*
the Surveillance and Control of Transmissible Animal Diseases, trans. from French edition, *Histoire de la Surveillance et du Contrôle des Maladies Animales,* 2000 (Paris:
Office International des Épizooties, 2003), 84; Lise Wilkinson, *Animals and Disease:*
An Introduction to the History of Comparative Medicine (Cambridge: Cambridge
University Press, 1992), 7.

    9. Blancou, *History of the Surveillance,* 84–85.

    10. Karl [Gottlieb] Haubner, *Handbuch der Veterinärpolizei* (Dresden: Schöfeld,
1869), 291.

    11. J[ean] Théodoridès, "Un Grand Médecin et Biologiste, Casimir-Joseph Da-
vaine (1812–1882)," *Analecta Medico-Historica,* vol. 4 (Oxford, UK: Pergamon Press,
1968), 69; he cites Jean-François Thomassin, *Dissertation sur la charbon malin de la*
Bourgogne, ou la pustule maligne (Dijon: Antoine Benoit, 1780). Miasmatic theories
linked illnesses to the environment, a particularly appropriate connection for anthrax given its persistence in the soil.

    12. Blancou, *History of the Surveillance,* 85.

    13. Louis Henri Joseph Hurtrel d'Arboval, *Dictionnaire de Médicine de Chirurgie*
et d'Hygiène Vétérinaire, 2nd ed., 6 vols. (Paris: J.-B. Baillière, 1838–1839), vol. I, 410.
My thanks to Jean Blancou for directing me to this source.

    14. O[tto] Bollinger, "Infectionen durch thierische Gifte: Milzbrand," *Hand-*
buch der Chronischen Infectionskrankheiten (Leipzig: Verlag von F.C. W. Vogel, 1874);
Bollinger, "Anthrax," in H. von Ziemssen, A. H. Buck, and G. L. Peabody, *Cyclope-*
dia of the Practice of Medicine (New York: William Wood, 1874–1881), vol. 3 (1875),
372–430, see 374.

    15. Carter, *The Rise of Causal Concepts of Disease,* 63. Carter elucidates the his-
tory of Western theorizing about disease causation in the nineteenth century.

    16. Of course, this brief description is a tremendous simplification.

    17. An excellent introduction to the intricacies of germ theories and practices is
Worboys, *Spreading Germs.* Although focusing on the UK, significant transnational
exchanges of ideas and practices make this study useful in general.

    18. See Gerald Geison, *The Private Science of Louis Pasteur* (Princeton, NJ:
Princeton, University Press, 1995); John Harley Warner, *Against the Spirit of the Sys-*
tem: The French Impulse in Nineteenth-Century American Medicine (Princeton, NJ:
Princeton University Press, 1998); Jessica Riskin, *Science in the Age of Sensibility: The*
Sentimental Empiricists of the French Enlightenment (Chicago: University of Chicago
Press, 2002); William Clark, Jan Golinski, and Simon Schaffer. *The Sciences in En-*
lightened Europe (Chicago: University of Chicago Press, 1999).

    19. See Caroline Hannaway, "Caring for the Constitution: Medical Planning in
Revolutionary France," *Transactions and Studies of the College of Physicians in Phila-*
delphia 14 (1992): 147–66; Laurence Brockliss and Colin Jones, *The Medical World of*
Early Modern France (New York: Clarendon Press, 1997); Charles Coulston Gillispie,

Science and Polity in France at the End of the Old Regime (Princeton, NJ: Princeton University Press, 1980).

20. Théodoridès, "Un Grand Médecin" 70.

21. Two historical works are recommended for more information about nineteenth-century experimental work on anthrax: William Bulloch, *The History of Bacteriology* (London: Oxford University Press, 1938), and Wilkinson, *Animals and Disease*. Histories published by the participants themselves include L. A. Raimbert, "Etude Historique sur le Charbon," *Gazette Médicin de Paris*, 3rd series, 22 (1867): 20–23, 58–61, 105–108, 135–39; Otto Bollinger, "Historisches über den Milzbrand und die stäbchenförmigen Körperchen," *Beitrage Vergl. Pathol. Path. Anat.* 2 (1872): 1–22; and Saturnin Arloing, F.-Isidore Cornevin, and M. Thomas, *Le Charbon Bacterien* (Paris: Montrou, 1882), chapter 1.

22. Jean Théodoridès, "Casimir Davaine (1812–1882): A Precursor to Pasteur," *Medical History* 10 (April 1966): 155–65, see 158.

23. Published as P[ierre]-F[rançois]-Olive R[ayer], "Inoculation du Sang de Rate," *Comptes Rendus des séances de la Société de Biologie et des ses Filiales* 2 (1850): 141–47; small bodies described on 142 and quotation on 144.

24. C[asimir]-J[oseph] Davaine, historical note, "Sur la découverte des bactéridies," *Bulletin Academie de Médicine* (Paris) 4 (1875): 581–84.

25. F[ranz] A[loys] A[ntoine] Pollender, "Mikroskopische un microchemische Untersuchung des Milzbrandblutes sowie über Wesen und Kur des Milzbrandes," *Vierteljahresschrift Gerichtlichen öff Medizin* 8 (1855): 103–14. See Bollinger, "Anthrax," 375.

26. Leon Z. Saunders, *Veterinary Pathology in Russia, 1860–1930* (Ithaca, NY: Cornell University Press, 1980), 27–29.

27. F. A. Brauell, "Versuche und Untersuchungen betreffend den Milzbrand des Menschen und der Theire," *Archiv fur Pathologische Anatomie und Physiologie und klinische Medizin* 11 (1857): 132–44.

28. F. A. Brauell, "Weitere Mittheilungen über Milzbrand und Milzbrandblut," *Archiv fur Pathologische Anatomie und Physiologie und klinische Medizin* 14 (1858): 432–66; Robert Koch, "Untersuchungen über Bacterian, V. Die Aetiologie der Milzbrand-Krankheit, begründet auf die Entwicklungsgeschichte des *Bacillus Anthracis*," *Beiträge zur Biologie der Pflanzen*, 2 (1876–1877): 277–310. The best English translation of this article, and the one used here, is Koch, "The Etiology of Anthrax, Founded on the Course of Development of the Bacillus Anthracis [1876]," *Essays of Robert Koch*, trans. K. Codell Carter, (New York: Greenwood Press, 1987), 1–17; see 11–12 for Koch's confirmation of Brauell's placental filter experiments.

29. Emile Duclaux, *Pasteur: The History of a Mind* (Philadelphia and London: W.B. Saunders, 1920), 234–35.

30. Bulloch, *History of Bacteriology*, 187.

31. Théodoridès, "Un Grand Médecin," 76.

32. H[enri] M[amer] O[nésine] Delafond, "Communication sur la Maladie Régnante," *Recueil de Médicine Véterinaire* 37 (1860): 726–48.

33. Louis Pasteur, "Animalcules infusoires vivant sans gaz oxygène libre et determinant des fermentations," *Comptes Rendus de l'Académie des Sciences* 52 (1861): 344–47; see also Pasteur, "Influence de l'oxygène sur le développement de la levure et la fermentation alcoolique," *Bulletin de la Société Chimique de Paris* (June 28, 1861): 79–82.

34. C[asimir] Davaine, "Recherches sur les infusoires du sang dans la maladie connue sous le nom de sang de rate," *Comptes Rendus de l'Académie des Sciences* 57 (1863): 220–23, quotation on 221.

35. Bulloch, *History of Bacteriology,* 187.

36. Monomorphism, or the theory that the bacilli had the same form regardless of host or location, underlay the important anthrax investigations of the next decades (chapter 3). See Susan D. Jones and Philip M. Teigen, "Anthrax in Transit: Practical Experience and Intellectual Exchange," *Isis* 99 (Sept. 2008): 455–85.

37. C. Davaine, "Recherches sur la nature at la constitution anatomique de la pustule maligne," *Comptes Rendus de l'Académie des Sciences* 60 (1865): 1296–99, quotations on 1296.

38. Davaine, "Recherches sur les infusoires," 222.

39. Hodges cited by Charles D. Homans, "Boston Society for Medical Improvement," *Boston Medical and Surgical Journal* 79 (Jan. 7, 1869): 359–61, see p. 360.

40. Théodoridès, "Casimir Davaine," 156.

41. Jacob Henle quoted in Thomas D. Brock, *Robert Koch: A Life in Medicine and Bacteriology* (Madison, WI: Science Tech Publishers, 1988), 28–29.

42. Ibid., 29–30.

43. Similar points are noted in ibid., 30–31.

44. Carter, "Introduction," *Essays of Robert Koch,* trans. K. Codell Carter (New York: Greenwood Press, 1987), ix–xxv, x.

45. Davaine's papers were not available to Koch, although Koch had read abstracts and reviews; see Koch, "Investigations of the Etiology of Wound Infections [1878]," *Essays of Robert Koch,* trans. K. Codell Carter (New York: Greenwood Press, 1987), 19–56, see 50. Like almost everyone else, Koch depended heavily on Otto Bollinger's essay, "Infectionen durch theirische Gifte," in Hugo Wilhelm von Ziemssen, *Handbuch der speciellen Pathologie und Therapie* (Leipzig: F.C.W. Vogel, 1874–1875), vol. 3, 464.

46. Brock, *Robert Koch,* 27; Carter, "Introduction," xi.

47. Bulloch, *History of Bacteriology,* 213.

48. Koch "The Etiology of Anthrax," 3–5.

49. Robert Koch, "Untersuchungen über Bacterian, V" 277. The best English translation of this article, and the one used here, is Koch, "The Etiology of Anthrax," 2. Dated December 26, 1876, this paper was followed eleven months later by "Untersuchungen über Bacterian, VI. Verfahren zur Untersuchung, zum Conserviren und Photographiren der Bacterien," *Beiträge zur Biologie der Pflanzen* 2 (1876–1877): 399–434.

50. Koch, "The Etiology of Anthrax," 3–4, 7.

51. Ibid., 9–10.

52. Ibid., 11.

53. Ibid., 13.

54. Brock, *Robert Koch,* 31.

55. Koch, "The Etiology of Anthrax," 2.

56. Ibid., 2, 3.

57. Ibid., 4, 5, 12.

58. Robert Koch, "Untersuchungen über Bacterian, V." In the interim, Koch, his editor (Ferdinand Cohn), or their printer decided to reproduce microphotographs without the middlemen necessary for representing bacteria in engravings or lithographs. "Untersuchungen V" included a lithograph of *B. anthracis,* drawn by hand on a stone by a craftsman; "Untersuchungen VI," however, included collotype plates representing Koch's microphotographs directly, made without intermediate drawings. John Burdon Sanderson probably had engravings drawn from the collotypes of "Untersuchnung VI" for his article in the *British Medical Journal* 1 (Feb. 9, 1878): 179–84, with figures on 181. My thanks to Philip M. Teigen for his expertise in this area.

59. Robert Koch, *Investigations into the Etiology of Traumatic Infective Diseases,* trans. W. Watson Cheyne (London: Syndenham Society, 1880), x.

60. Koch, *Traumatic Infective Diseases.* On the significance of photographs as authoritative visual representations, see David N. Livingstone, *Putting Science in Its Place: Geographies of Scientific Knowledge* (Chicago: University of Chicago Press, 2003), 163–71.

61. John Burdon Sanderson, "Lectures on the Infective Processes of Disease," *British Medical Journal* 1 (Feb. 9, 1878): 179–84, quotation in caption to figures on 181.

62. The disputes owed much to French-German nationalism. As K. Codell Carter has concluded, both Pasteur and Koch made major contributions to establishing causation for *B. anthracis.* See Carter, "The Koch-Pasteur Dispute on Establishing the Cause of Anthrax," *Bulletin of the History of Medicine* 62 (1988): 42–57.

63. Edwin Klebs, "Ueber die Umgestaltung der medicinischen Anschauungen in den letzten drei Jahrzehnten," *Tageblatt der 50 Versammlung der Gesellschaft deutscher Naturforscher und Aerzte* (Munich: R. Straub, 1877), 41–55; Bulloch, *History of Bacteriology,* 219.

64. Joseph Lister, "On Lactic Fermentation," *Transactions of the Pathological Society* (London) 29 (1878): 425–67; Bulloch, *History of Bacteriology,* 222–23.

65. Bulloch, *History of Bacteriology,* 224–25.

66. Robert Koch, *Untersuchungen über die Aetiologie der Wundinfectionskrankheiten* (Leipzig: Georg Thieme, 1878). The best English translation of this essay, and the one used here, is "Investigations of the Etiology of Wound Infections [1878]," *Essays of Robert Koch,* trans. K. Codell Carter (New York: Greenwood Press, 1987), 19–56, see 51.

67. Koch "Investigations of the Etiology of Wound Infections," 51.

68. Louis Pasteur and J[ules] F[rançois] Joubert, "Étude sur la maladie charbon-neuse," *Comptes Rendus des séances de la Société de Biologie* 84 (1877): 900–906. These authors went through one hundred inoculations of the bacillus from one animal to another and still produced the disease in the last animal. They concluded that this, and not another bacillus, was inextricably associated with anthrax.

69. W. S. Greenfield [communicated by J. Burdon-Sanderson], "Preliminary Note on some Points in the Pathology of Anthrax, with especial reference to the Modification of the properties of the Bacillus anthracis by Cultivation, and to the Protective Influence of Inoculation with a Modified Virus," *Proceedings of the Royal Society London* 30 (June 17, 1880): 557–60, see 557; J. B. Sanderson, "Report on Experiments on Anthrax Conducted at the Brown Institution, February 18 to June 30, 1878," *Journal of the Royal Agricultural Society* 41 (1880): 267–72.

70. Greenfield, "Preliminary Note," 558.

71. Ibid., 559–60.

72. W. S. Greenfield, "Abstracts of Lectures on Further Investigations on Anthrax and Allied Diseases in Man and Animals," *British Medical Journal* 2 (Dec. 25, 1880): 1007–1009; 1 (Jan. 1, 1881): 3–5, 81–82. For some context surrounding these lectures and the Brown Institution, see Wilkinson, *Animals and Disease,* 172–74; Terrie M. Romano, *Making Medicine Scientific: John Burdon Sanderson and the Culture of Victorian Science* (Baltimore, MD: Johns Hopkins University Press, 2002), esp. chapter 5.

73. Worboys, *Spreading Germs,* 65.

74. Geison, *The Private Science of Louis Pasteur,* 160–61.

75. Henri Toussaint, "Note Contenue dans un pli cacheté et relative à un Procédé pour la Vaccination du Mouton et du Jeune Chien," *Comptes Rendus de l'Académie des Sciences* 91 (Aug. 2, 1880): 415–18; Théodoridès, "Un Grand Médecin," 119, 145; Charles-Ernest Cornevin, "Nécrologie. MH Toussaint," *Journal Médecin Véterinaire Zootechnologie,* 3rd series, 14 (Aug. 1890): 438–41.

76. "Jean-Joseph Henri Toussaint," obituary, *Lyon Médicin* 65 (1890): 55–65. Toussaint's most important paper was "De l'immunité pour le charbon acquise à la suite d'inoculations préventives," *Comptes Rendus de l'Académie des Sciences* 91 (1880): 135–37.

77. Charles Chamberland and Emile Roux, "Sur l'attenuation de la virulence de la bacteridie charbonneuses, sous l'influence des substances antiseptiques," *Comptes Rendus de l'Académie des Sciences* 96 (1883): 1088–91.

78. Geison, *The Private Science of Louis Pasteur.*

79. Ibid., 145, 167.

80. Latour, *The Pasteurization of France,* 87–90, quotation on 87.

81. Geison, *The Private Science of Louis Pasteur,* 151.

82. Ibid., 167–68.

83. Ibid., 176.

84. Maurice Cassier, "Appropriation and Commercialisation of the Pasteur Anthrax Vaccine," *Studies in the History and Philosophy of Biological and Biomedical Sciences* 36 (2005): 722–42, see 723.

85. Ibid., 724.

86. Anne-Marie Moulin, "Patriarchal Science: The Network of the Overseas Pasteur Institutes," in *Science and Empires: Historical Studies about Scientific Development and European Expansion,* ed. Patrick Petitjean, Catherine Jami, and Anne-Marie Moulin (Dordrecht: Kluwer Academic, 1992), 307–22; Jean-Pierre Dedet, *Les Instituts Pasteur d'Outre-Mer: Cent Vingt Ans de Microbiologie Française dans le Monde* (Paris: L'Harmattan, 2000). Thanks to Tamara Giles-Vernick for directing me to these studies.

87. Cassier, "Appropriation and Commercialisation," 736–37; Jean Chaussivert and Maurice Blackman, eds., *Louis Pasteur and the Pasteur Institute in Australia: Papers from a Symposium Held at the University of New South Wales, 4–5 September 1987* (Kensington: French-Australian Research Centre, University of New South Wales, 1988), 25–37.

88. Bouley, "La nouvelle vaccination."

89. Worboys, *Spreading Germs,* 194.

90. Edward Tibbets, "Woolsorters' Disease," Correspondence, *British Medical Journal* 1 (Feb. 19, 1881): 293.

91. Greenfield did reply to a similar query from Tibbets in the *Lancet* (vol. 1, Feb. 12, 1881, 276), in which he scolded Tibbets for not reading the reports of Greenfield's lectures carefully enough.

92. Leonard Cane, "Woolsorters' Disease," Correspondence, *British Medical Journal* 1 (Feb. 26, 1881): 325.

93. Edward T. Tibbits (sic), "Woolsorters' Disease," Correspondence, *British Medical Journal* 1 (June 18, 1881): 982.

94. Jones and Teigen, "Anthrax in Transit."

95. Worboys, *Spreading Germs,* 140.

96. Theodor Rosebury, *Peace or Pestilence? Biological Warfare and How to Avoid It* (New York: McGraw-Hill, 1949), 68.

Chapter 3. Transmission

Epigraph: J. H. Bell, "On "Woolsorter's Disease," *Lancet* 1 (1880): 871–73, quotation on 872.

1. Jack Reynolds, *The Great Paternalist: Titus Salt and the Growth of Nineteenth-Century Bradford* (London: Maurice Temple Smith; New York: St Martin's Press, 1983), 52–53. According to the *Encyclopedia Brittanica,* 1911, the importation of fiber from Turkey, Egypt, and North Africa tripled between 1840 and 1860, and again between 1860 and 1880.

2. John F. Richards, "Early Modern India and World History," *Journal of World History* 8, 2 (Fall 1997): 197–209, see 199.

3. Susan M. Ouellette, *U.S. Textile Production in Historical Perspective: A Case Study from Massachusetts* (New Brunswick, NJ: Routledge, 2007); Jack Reynolds, *The Great Paternalist;* D. T. Jenkins, ed., *The Textile Industries* (Oxford; Cambridge, MA: Blackwell, 1994); Howard Spodek, *The World's History Since 1100* (Upper Saddle River, NJ: Prentice Hall, 2001); William J. Duiker, *World History* (Belmont, CA: Thomson/Wadsworth, 2004); William H. McNeill, *A World History* (New York: Oxford University Press, 1999).

4. B. W. de Vries, *From Peddlers to Textile Barons: The Economic Development of a Jewish Minority Group in the Netherlands* (Amsterdam: North Holland, 1989); Carmen Ramos-Escandon, *Industrializacíon, Genero y Trabajo Feminino en el Sector Textil Mexicano: El Obraje, la Fabrica y la Compania Industrial* (Mexico, D.F.: CIESAS, 2004).

5. Tony Jowitt, "The Retardation of Trade Unionism in the Yorkshire Worsted Textile Industry," in *Employers and Labour in the English Textile Industries,1850–1939,* ed. J. A. Jowitt and A. J. McIvor (London and New York: Routledge, 1988), 84–106, see 86; R. G. Wilson, "The Supremacy of the Yorkshire Cloth Industry in the Eighteenth Century," in D. T. Jenkins, ed., *The Textile Industries,* vol. 8 of *The Industrial Revolutions,* ed. R. A. Church and E. A. Wrigley, (Oxford: Blackwell, 1994), 225–46.

6. Ramos-Escandon, *Industrializacíon, Genero y Trabajo Feminino.*

7. Paul E. Rivard, *A New Order of Things: How the Textile Industry Transformed New England* (Hanover, NH: University Press of New England, 2002), 15–18, 80, 83, 92–94.

8. "Wool, Worsted and Woollen Manufactures," *Encyclopedia Britannica,* 1911 edition, table A, Turkish, Egyptian and North African row, www.1911encyclopedia.org/Wool,_worsted_and_woollen_ manufactures (accessed Aug. 14, 2007).

9. "Mohair," *Encyclopedia Britannica,* 1911 edition, www.1911encyclopedia.org/Mohair (accessed on Aug. 14, 2007).

10. John Spear, "On the So-Called 'Woolsorters' Disease' as Observed at Bradford and in Neighbouring Districts in the West Riding of Yorkshire," appendix A, no. 8, *Tenth Annual Report of the Local Government Board, 1880–81: Supplement containing the Report of the Medical Officer for 1880* (London: George E. Eyre and William Spottiswoode, for Her Majesty's Stationery Office, 1881), 66–135; see 85.

11. "Wool, Worsted, and Woollen Manufactures," *Encyclopedia Britannica,* 1911 edition table A, Turkish, Egyptian and North African row, www.1911encyclopedia.org/Wool,_worsted_and_woollen_ manufactures (accessed on Aug. 14, 2007).

12. Nicolas Fournier, *Observations et experiences sur les charbons malins* (Dijon, 1757 and 1769); David M. Morens, "Characterizing a 'New' Disease: Epizootic and Epidemic Anthrax, 1769–1780," *American Journal of Public Health* 93 (June 2003): 886–93.

13. Rudolf L. K. Virchow, "Milzbrand und Karbunkelkrankheit," in *Intoxicationen, Zoonosen und Syphilis, Handbook der Speciellen Pathologie und Therapie,* vol. 2, pt. 1, ed. C. Ph. Falck, Rudolf L. K. Virchow, and F. A. Simon (Erlangen: Ferdinand Enke, 1855), 387–405; Bollinger, "Anthrax," 407.

14. C[asper] W. Pennock, "On the Malignant Pustule; with cases," *American*

Journal of the Medical Sciences 37 (Nov. 1836): 13–25. Pennock cites Alexis Boyer, *Traite des Maladies Chirurgicales et des Operations qui leer Conviennent,* 11 vols. (Paris: The Author and Migneret, 1814–1826). Boyer (1757–1833) was a surgeon to Napoleon and succeeding French monarchs.

15. J[oseph] M[aximilian] Chelius, *Handbuch der Chirurgie: zum Gebrauche bei seinen Vorlesungen / Sechste, vermehrte und verbesserte Original-Auflage* (Heidelberg and Leipzig: Neue Akdemische Buchandlung von Karl Groos; Wien: bei Karl Gerold, 1843); an English translation by John F. South, *A System of Surgery, by J.M. Chelius,* was printed in London and in Philadelphia (Lea and Blanchard) in 1847.

16. O[tto] Bollinger, "Infectionen durch thierische Gifte: Milzbrand," *Handbuch der Chronischen Infectionskrankheiten* (Leipzig: Verlag von F.C. W. Vogel, 1874); Bollinger, "Anthrax," in H. von Ziemssen, A. H. Buck, and G. L. Peabody, *Cyclopedia of the Practice of Medicine* (New York: William Wood, 1874–1881), vol. 3 (1875), 372–430, see 409–10.

17. Although it should be noted that anthrax was not a notifiable disease in animals in Great Britain until 1886 (and even later in the United States).

18. Pennock, "On the Malignant Pustule," 14.

19. William Budd, "Observations on the Occurrence of Malignant Postule in England: Illustrated by Numerous Fatal Cases," *British Medical Journal* 1 (Jan. 24, 1863): 85–87. P. W. J. Bartrip, *The Home Office and the Dangerous Trades: Regulating Occupational Disease in Victorian and Edwardian Britain* (Amsterdam and New York: Rodopi, 2002), 235. Bartrip gives less credibility to the argument that industrialization and global trade increased the incidence of anthrax in the mills.

20. F[rederick] W. Eurich, "The History of Anthrax in the Wool Industry of Bradford, and of Its Control," *Lancet* 1 (Jan. 2, 1926): 57–58, 107–109, see 57.

21. *Bradford Observer,* Jan. 4, 1855, 4.

22. Reynolds, *The Great Paternalist,* 310.

23. Pennock, "On the Malignant Pustule," 13.

24. P. S. Brachman, H. Gold, S. A. Plotkin, F. R. Rekety, M. Werrin, and N. R. Ingraham, "Field Evaluation of a Human Anthrax Vaccine," *American Journal of Public Health* 52 (1962): 632–45.

25. I am indebted to Philip M. Teigen for drawing my attention to this outbreak, supplying much of the source material, and engaging in stimulating discussions on the topic.

26. Silas E. Stone, "Malignant Pustule—Charbon Fever," *Publications of the Massachusetts Medical Society,* vol. III (Boston: David Clapp and Son, 1872), 83–98. See 88–89 for the description of Sarah R.'s illness, and 84 for the tally of cases.

27. "Dangers of Curled Hair: Deaths from Malignant Pustule among the Operatives in Glover and Whitcomb's Factory at Hyde Park," *Boston Daily Advertiser,* Aug. 27, 1878, col. A.

28. "Charbon Fever: Disease and Death in a Hair Mattress Manufactory," *Boston Globe,* Aug. 27, 1878 (morning edition).

29. Ibid.

30. Silas E. Stone, "Cases of Malignant Pustule," *BMSJ* 78 [n.s. 1] (Feb. 13, 1868): 19–21. This article has also been described in Abe Macher, "Industry-Related Outbreak of Human Anthrax, Massachusetts, 1868," letter, *Emerging Infectious Diseases* 8 (Oct. 2002): 1182.

31. Stone, "Cases," 21; see Pierre Rayer, *A Dictionary of Practical Medicine* (New York: Harper and Brothers, 1852): 612–15.

32. For more on Parisian scientific and medical culture, see John Harley Warner, *Against the Spirit of the System: The French Impulse in Nineteenth-Century American Medicine* (Princeton, NJ: Princeton University Press, 1998).

33. Stone, "Cases," 21.

34. Silas E. Stone, "Malignant Vesicle," *BMSJ* 80 [n.s. 3] (Feb. 11, 1869): 21–24.

35. Hodges had studied midwifery in Dublin and anatomy and surgery in Paris. He also served as demonstrator of anatomy under Harvard's Oliver Wendell Holmes. See Walter L. Burrage, "Richard Manning Hodges," *Dictionary of American Medical Biography* (Boston: Milford House, 1978, reprint from 1928 edition), 576.

36. Casimir Davaine, "Recherches sur les infusoires du sang dans la maladie connue sous le nom de sang de rate," *Comptes Rendus de l'Académie des Sciences* 57 (1863): 220–23.

37. Charles D. Homans, "Boston Society for Medical Improvement," *BMSJ* 79 [n.s. 2] (Jan. 7, 1869): 359–61, see 360. Homans was the secretary of the society. According to Harrington, Homans and Hodges had been in Paris together.

38. Toby A. Appel, "A Scientific Career in the Age of Character: Jeffries Wyman and Natural History at Harvard," in *Science at Harvard University: Historical Perspectives,* ed. Clark A. Elliott and Margaret W. Rossiter (Bethlehem, PA: Lehigh University Press, 1992), 96–120, see esp. page 98 and quotation on 112.

39. Stone, "Malignant Vesicle," 23.

40. Homans, "Boston Society for Medical Improvement," 360.

41. Susan D. Jones and Philip M. Teigen, "Anthrax in Transit: Practical Experience and Intellectual Exchange," *Isis* 99 (Sept. 2008): 455–85; James Strick, *Sparks of Life: Darwinism and the Victorian Debates over Spontaneous Generation* (Cambridge, MA: Harvard University Press, 2000), 50–60.

42. See, for example, David Russell, *Reminiscences of My Father* (Dunfries: Thomas Hunter, privately published ,1906), held by the British Library. David Russell recounted a compliment once paid to his father by John Shaw Billings of New York, who said that James Russell owned "one of the best working libraries he had ever come across" (10). Billings was librarian at the New York Public Library and well known for his work at the Surgeon General's Library (*Index Medicus,* etc.) and in public health. For more on Russell, see Edna Robertson, *Glasgow's Doctor: James Burn Russell, 1837–1904* (East Lothian, UK: Tuckwell Press, 1998).

43. James Burn Russell, "On the Contagium of Anthrax to Hair Factory Workers, as Illustrative of the Particulate Theory of the Contagia," *Report of the Medical*

Officer Local Government Board (London: HMSO, 1879), appendix 7, 321–45. Russell also read an augmented version of this paper as an address to the Philosophical Society of Glasgow on April 28, 1880, which published it; the address was also published in the *Sanitary Journal* (Glasgow), no. 52 (June 1, 1880): 97–120.

44. Russell, "On the Contagium," 8–9.

45. Ibid., 19.

46. For a published version of this lecture, see John Tyndall, "Fermentation and its Bearing on the Phenomenon of Disease," *Fortnightly Review* 20 (1876): 567. Tyndall learned of Koch's work from a visit by Ferdinand Cohn, who had seen Koch's demonstration of anthrax experiments back in Breslau (prior to the publication of Koch's paper). See Thomas D. Brock, *Robert Koch: A Life in Medicine and Bacteriology* (Madison, WI: Science Tech Publishers, 1988), 50–51.

47. James Christie, *The Medical Institutions of Glasgow* (Glasgow: James Maclehose, 1889), 198.

48. James B. Russell, "On the Conveyance of the Contagium of Anthrax to Hair Factory Workers, as illustrative of the Particulate theory of the Contagia," *Proceedings of the Philosophical Society of Glasgow* (Glasgow: Philosophical Society, 1880), 3, 8. He referred to his article "On the Contagium," which was appended to Edward Ballard, "Third Report in Respect of the Inquiry as to Effluvium Nuisances Arising in Connexion with Various Manufacturing and Other Branches of Industry," *Report of the Medical Officer Local Government Board* (London: HMSO, 1879), 42–48

49. Russell, "On the Contagium," 25.

50. Reynolds, *The Great Paternalist*, 52–53.

51. Bell's intensive public opinion campaign, waged in the *Bradford Observer*, began with a letter to the editor about a worker's death, "Deaths from Blood Poisoning in Bradford," published on February 26, 1878, p. 4. The *Observer* was apparently quite sympathetic, because it devoted substantial space to details of autopsy reports and epidemiological investigations over the next few weeks.

52. Bradford woolsorters had a history of agitation on the issue even before Bell's arrival in the area in the early 1860s. See, for example, *Bradford Observer*, Jan. 4, 1855, and *Reynolds's Newspaper*, Nov. 1, 1857 (n.p.).

53. John Henry Bell, notebook, DB15 C5, West Yorkshire Archives, Bradford Central Library, Bradford, UK, n.p.

54. Ibid.

55. "Rules and Orders of the Bradford Woolsorters' Society" (Bradford: John Dale, 1839), bound together with "Rules and Orders of the Bradford Woolsorters' Society" (Bradford: John Dale, 1840), box 3-70, Empsall Collection, Local History Collection, Bradford Central Library, Bradford, UK; Reynolds, *The Great Paternalist*, 32–34, 158, 333.

56. "Rules and Orders of the Bradford Woolsorters' Society," 1839. The fines were sixpence or more.

57. Minutes of the Bradford Medico-Chirurgical Society, Feb. 5, 1878, 40/D/89/2, West Yorkshire Archives, Bradford, UK.

58. For a brief biography of Eddison, see S. T. Anning, *The General Infirmary at Leeds,* vol. 2, *The Second Hundred Years* (Edinburgh and London: E. and S. Livingstone, 1966), 131. Eddison was part of the intellectual elite of Leeds; he was also president of the Leeds Philosophical Society in 1883–1885. John Henry Bell, "On 'Woolsorter's Disease,'" *Lancet* 1 (June 5, 1880): 871–73, 909–11, see 911; Samuel Lodge, "La Maladie des Triers de Laine," *Archives de Medicine Experimentale* 2 (1890): 759–71, see 760. Bell was quite active in the Leeds society, serving on its board from 1876 to 1880; Eddison joined him on the board in 1878 and 1879. See *Medical Directory* (London: J. A. Churchill) for 1876 (p. 746), 1877 (p. 769), 1878 (p. 771), 1879 (p. 789), and 1880 (p. 813).

59. This was listed as Case 5 in Spear's report, a man named J. W. Spear, "On the So-Called 'Woolsorters' Disease.'"

60. Elizabeth Fee and T. M. Brown, "John Henry Bell: Occupational Anthrax Pioneer," *American Journal of Public Health* (2002) 92: 756–57.

61. Bartrip, *The Home Office and the Dangerous Trades,* 239.

62. See *British Medical Journal* 2 (July 10, 1880): 54; 2 (July 24, 1880): 139.

63. W. S. Greenfield, "Abstracts of Lectures on Further Investigations on Anthrax and Allied Diseases in Man and Animals," *British Medical Journal* 2 (Dec. 25, 1880): 1007–1009; 1 (Jan. 1, 1881): 3–5, 81–82. For some context surrounding these lectures and the Brown Institution, see Lise Wilkinson, *Animals and Disease: An Introduction to the History of Comparative Medicine* (Cambridge: Cambridge University Press, 1992), 172–74; Terrie M. Romano, *Making Medicine Scientific: John Burdon Sanderson and the Culture of Victorian Science* (Baltimore, MD: Johns Hopkins University Press, 2002), esp. ch. 5.

64. Spear, "On the So-Called 'Woolsorters' Disease.'" 96.

65. Ibid., 97.

66. Ibid., 113, 134–35, quotations on 135. The fact that fewer workers died of dallack may have been due to underreporting; but these workers also did their sorting and bundling outdoors (better ventilation decreased the number of spores they inhaled) and may have had some immunity through long exposure.

67. Matthew N. Van Ert, W. Ryan Easterday, Lynn Y. Huynh, et al., "Global Genetic Population Structure of *Bacillus anthracis,*" *PLoS One* 2, 5 (May 2007): e461, 1–10.

68. Ibid., 6.

69. Ibid., 7.

70. Ibid.

71. Ibid.

72. Ian Mortimer and Joseph Melling have characterized the responses to anthrax as technical, medical, and the "Factory Act model." See "'The Contest be-

tween Commerce and Trade, on the One Side, and Human Life on the Other': British Government Policies for the Regulation of Anthrax Infection and the Wool Textiles Industries, 1880–1939," *Textile History* 31, 2 (2000): 222–36, see 232.

73. Bradford Medico-Chirurgical Society, *Report of the Commission on Wooolsorters'* [sic] *Diseases,* (Bradford: Toothill, 1882), B614.561 BRA and Ref.194.4/439, Local Collections, Bradford Central Library, Bradford, Yorkshire.

74. Ibid., 16.

75. A copy of this clipping may be found in scrapbook 1, J. H. Bell and F. W. Eurich, "Papers on Anthrax Collected by Dr J. H. Bell and Dr Fritz Eurich, 1878–1911," GB 0532, Bradford University Library, Bradford, Yorkshire.

76. F. W. Eurich cited Bell in his 1926 lecture before the Bradford Medico-Chirurgical Society [published as F.W. Eurich, "The History of Anthrax in the Wool Industry of Bradford, and of Its Control," *Lancet* i (Jan. 2, 1926): 57–58, see 58].

77. Reynolds, *The Great Paternalist,* 349.

78. Mortimer and Melling, "'The Contest between Commerce and Trade,'" 224.

79. See Legge's excellent Milroy Lectures to the Royal College of Physicians (1905). T. M. Legge, "Industrial Anthrax," *Lancet* 1 (March 18, March 25, April 1, 1905): 689–94, 765–76, 841–45.

80. See the *Annual Reports, 1906–1918, Anthrax Investigation Board for Bradford and District,* B614.561 ANT, Local Collections, Bradford Central Library, Bradford, Yorkshire; J. H. Bell and F. W. Eurich, "Papers on Anthrax"; F.W. Eurich, "The History of Anthrax in the Wool Industry."

81. Tim Carter, "Anthrax: A Global Problem with an Italian Cure," in *Contributions to the History of Occupational and Environmental Protection: 1st International Conference on the History of Occupational and Environmental Prevention, Rome, Italy, 1998,* ed. Antonio Grieco, Sergio Iavicoli, and Giovanni Berlinguer (Amsterdam: Elsevier, 1999), 247–51.

82. For a more detailed description of these events, see Mortimer and Melling, "'The Contest between Commerce and Trade,'" 226–27.

83. Ibid., 228.

84. N. Metcalfe, "The History of Woolsorter's Disease: A Yorkshire Beginning with an International Future?" *Occupational Medicine* 54 (2004): 489–93, see 492.

85. U.S. officials claimed that the British disinfection plant had caused the diversion of much potentially dangerous wool to the United States, leading to an increase in anthrax cases in workers. See "Anthrax in the United States," *Monthly Labor Review* 51 (Oct. 1940): 946–50, see 948.

86. Bollinger, "Anthrax," 403.

87. Karl [Gottlieb] Haubner, *Handbuch der Veterinär-Polizei* (Dresden: Schönfeld, 1869), 291.

88. Bollinger, "Anthrax," 404.

89. C[ecil] H. W. Page, "British Industrial Anthrax," *Journal of Hygiene* 9 (1909): 279–315, see 293–94.

90. For a valuable summary of published case reports in twentieth-century Germany and other nations, see Jon-Erik Holty, Dena M. Bravata, Hau Liu, Richard A. Olshen, Kathryn M. McDonald, and Douglas K. Ownes, "Systematic Review: A Century of Inhalational Anthrax Cases from 1900 to 2005," *Annals of Internal Medicine* 144, 4 (2006): 270–80.

91. Daniel Gilfoyle, "Anthrax in South Africa: Economics, Experiment and the Mass Vaccination of Animals, c. 1910–1945," *Medical History* 50 (2006): 465–90, see 472.

92. Maurice Cassier, "Appropriation and Commercialisation of the Pasteur Anthrax Vaccine," *Studies in the History and Philosophy of Biological and Biomedical Sciences* 36 (2005): 722–42, see 723.

93. Ibid., 724.

94. Anne-Marie Moulin, "Patriarchal Science: The Network of the Overseas Pasteur Institutes," in *Science and Empires: Historical Studies about Scientific Development and European Expansion,* ed. Patrick Petitjean, Catherine Jami, and Anne-Marie Moulin (Dordrecht: Kluwer Academic, 1992); Jean-Pierre Dedet, *Les Instituts Pasteur d'Outre-Mer: Cent Vingt Ans de Microbiologie Française dans le Monde* (Paris: L'Harmattan, 2000). Thanks to Tamara Giles-Vernick for directing me to these studies.

95. Cassier, "Appropriation and Commercialisation," 736–37; Jean Chaussivert and Maurice Blackman, eds., *Louis Pasteur and the Pasteur Institute in Australia: Papers from a Symposium Held at the University of New South Wales, 4–5 September, 1987* (Kensington, N.S.W.: French-Australian Research Centre, University of New South Wales, 1988), 25–37.

96. Adolphe Debray, *Le Charbon Industriel: Maladie ou Accident Professionnels* (Paris: Bonvalot-Jouve, 1906).

97. For the case of tuberculosis, see Susan D. Jones, "Mapping a Zoonotic Disease: Anglo-American Efforts to Control Bovine Tuberculosis Before World War I," *Osiris* 19 (2004): 133–48.

98. Gilfoyle, "Anthrax in South Africa," 479.

99. For animal diseases, see Keir Waddington, "To Stamp Out 'So Terrible a Malady': Bovine Tuberculosis and Tuberculin Testing in Britain, 1890–1939," *Medical History* 48 (2004): 29–48; Susan D. Jones, *Valuing Animals: Veterinarians and Their Patients in Modern America* (Baltimore, MD: Johns Hopkins University Press, 2003), 108–109; Gregg Mitman, *Breathing Space: How Allergies Shape Our Lives and Landscapes* (New Haven, CT: Yale University Press, 2007), 120–21, 206–50.

100. Emile Marchoux, "Serum Anticharbonnuex," *Annales de l'Institut Pasteur* 9 (Nov. 1895): 800–806; Joseph C. Regan, "The Advantage of Serum Therapy as Shown by a Comparison of Various Methods of Treatment of Anthrax," *American Journal of the Medical Sciences* 162 (Sept. 1921): 407–23. A German scientist, G. Sobernheim, developed yet another method at about the same time.

101. Page, "British Industrial Anthrax," 291. Mandatory notifications and better

reporting contributed to these higher statistics, especially compared with France and eastern European nations.

102. See T[homas] M. Legge, "Report on Incidences of Anthrax in Manipulation of Horsehair and Bristles" (London: HMSO, 1906), 69.

103. Adrianus Pijper, "The Treatment of Human Anthrax," *Lancet* 1 (Jan. 9, 1926): 88–89.

104. Regan, "Various Methods of Treatment of Anthrax," 413. See Regan's references for the wide use of the serum.

105. R. H. Boggon, "Four Cases of Anthrax Treated with Sclavo's Serum," *Lancet* 1 (Feb. 26, 1927): 435–36; Regan, "Various Methods of Treatment of Anthrax," 411–20.

106. A[dolph] Eichhorn, "Vaccination Experiments Against Anthrax," *Journal of the American Veterinary Medical Association* 47 (March 1916): 669–87, see 669; and ibid. (Oct., 1915): 1479 (correspondence).

107. Pijpers, "Treatment of Human Anthrax," 89.

108. Louisiana Veterinary Medical Association, *Veterinary Medicine in Louisiana: 1889 to 1979,* Hornsby Edition ([Louisiana]: Louisiana Veterinary Medical Association, 1993), 128.

109. Committee on Industrial Anthrax, "Cases of Industrial Anthrax, 1929 to 1933," *American Public Health Association Yearbook* (New York: American Public Health Association, 1935), 1240–42; Committee on Anthrax, "Anthrax in the United States, 1919 to 1938," *Monthly Labor Review* 51 (Oct. 1940): 946–50.

110. Max Sterne, "Variation in *Bacillus anthracis,*" *Onderstepoort Journal of Veterinary Science and Animal Industry* 8 (1937): 271–350, see 286–87.

111. J. H. Mason, "A New Culture Tube," *Journal of the South African Veterinary Medical Association* 4 (1933): 89–90.

112. Gilfoyle, "Anthrax in South Africa," 486–88.

113. Max Sterne, "Avirulent Anthrax Vaccine," *Onderstepoort Journal of Veterinary Science and Animal Industry* 21, 1 (1946): 41–43.

114. Gilfoyle, "Anthrax in South Africa," 465.

115. Peter Turnbull, "Obituary: Max Sterne," *Independent,* March 4, 1997.

116. The Sclavo name is still attached to the company, which manufactures medical diagnostic products. Alison Abbott, "A Place in the Sun," *Nature* 446 (March 8, 2007): 124–25.

117. Health Protection Agency, "History of Anthrax Vaccine Manufacture," www.hpa.org.uk/web/HPAwebFile/HPAweb_C/1194947412710 (accessed Jan. 29, 2010).

118. See for example G. G. Wright and J. B. Slein, "Studies on Immunity in Anthrax, I. Variation in the Serum T-Agglutinin During Anthrax Infection in the Rabbit," *Journal of Experimental Medicine* 93 (1951): 99–102; G. G. Wright, T. W. Green, and R. G. Kanodo, "Studies on Immunity in Anthrax, V. Immunizing Activity of Alum-Precipitated Protective Antigen," *Journal of Immunology* 73 (1955): 387–98.

119. G. P. Gladstone, "Immunity to Anthrax: Protective Antigen Present in Cell-Free Culture Filtrates," *British Journal of Experimental Pathology* 27 (1946): 394–418; F. C. Belton and R. E. Strange, "Studies on a Protective Antigen Produced *in vitro* from *Bacillus anthracis:* Medium and Methods of Production," *British Journal of Experimental Pathology* 35 (1954): 144–49.

120. H. M. Darlow, F. C. Belton, and D. W. Henderson, "The Use of Anthrax Antigen to Immunise Man and Monkey," *Lancet* 2 (Sept. 8, 1956): 476–79.

121. The Manchester mill was one of four involved in these clinical trials. See Brachman et al., "Field Evaluation." Quote from Peter C. B. Turnbull, "Anthrax Vaccines: Past, Present and Future," *Vaccine* 9 (Aug. 1991): 533–39, see 535. This article is an excellent introduction to the history of anthrax vaccine development.

122. Richard M. Swiderski, *Anthrax: A History* (Jefferson, NC: McFarland, 2004), 33–43.

123. Robert C. Simpson, "Anthrax," *American Journal of Nursing* 48, 12 (Dec. 1948): 759–61, see 760.

124. H. Gold, "Anthrax: A Report of 117 Cases," *American Medical Association Archives of Internal Medicine* 96 (1955): 387–96.

125. Bartrip, *The Home Office and the Dangerous Trades,* 249; Mortimer and Melling, "'The Contest between Commerce and Trade,'" 230–32.

126. Worker efficiency was a common rhetorical, if not actual, motivation behind many antidisease campaigns in this period (tuberculosis being the most notable example).

127. Bradford Medico-Chirurgical Society, *Report of the Commission on Wooolsorters'* [sic] *Disease,* 12.

128. Mortimer and Melling, "'The Contest between Commerce and Trade,'" 233.

129. Jowett's words are recorded in *Hansard* 5, 41 (July 17, 1912): 464.

130. Bartrip, *The Home Office and the Dangerous Trades,* 251.

131. Owsei Temkin, "The Double Face of Janus," *The Double Face of Janus and Other Essays in the History of Medicine* (Baltimore, MD: Johns Hopkins University Press, 1977), 3–37, quotation on 28–29.

Chapter 4. Casualty Effectiveness

Epigraph: *Development of 'N' for Offensive Use in Biological Warfare: Special Report No. 9,* Detrick, July 1944, RG 160, Record of the Army Service Forces, National Archives Records Administration, Washington, D.C., also cited in John Ellis van Courtland Moon, "U.S. Biological Warfare Planning and Preparedness: The Dilemmas of Policy," in *Biological and Toxin Weapons: Research, Development and Use from the Middle Ages to 1945,* ed. E. Geissler and J. E. van Courtland Moon, Stockholm international Peace Research Institute, Chemical and Biological Warfare Studies No. 18 (Oxford: Oxford University Press, 1999), 215–54, see 249.

1. Valentin Bojtzov and Erhard Geissler, "Military Biology in the USSR, 1920–45," *Biological and Toxin Weapons,* 153–67, see 157.

2. The term "casualty effectiveness" appeared in this context in Theodor Rosebury and Elvin A. Kabat, "Bacterial Warfare: A Critical Analysis of the Available Agents, Their Possible Military Applications, and the Means for Protection Against Them," *Journal of Immunology* 56, 1 (May 1947): 7–96. Rosebury and Kabat originally wrote this report in 1942, but it was published only after the war.

3. Leon A. Fox, "Bacterial Warfare: The Use of Biologic Agents in Warfare," *Military Surgeon* 72, 3 (1933): 189–207.

4. Jeanne Guillemin, *Biological Weapons: From the Invention of State-Sponsored Programs to Contemporary Bioterrorism* (New York: Columbia University Press, 2005), 4–6, 116.

5. Theodor Rosebury, *Peace or Pestilence: Biological Warfare and How to Avoid It* (New York: McGraw-Hill, 1949), 36.

6. William M. Creasy, "Biological Warfare," *Armed Forces Chemical Journal* 5 (Jan. 1952): 16–18, 46, see 18. The phrase was used again in *Effects of Biological Warfare Agents*, a pamphlet published by the U.S. Department of Health, Education, and Welfare in July 1959. It appears often in the open literature from the 1960s onward to condemn chemical and biological weapons research as distortions of scientific values and practices.

7. See Brian Balmer, "Killing 'Without The Distressing Preliminaries': Scientists' Defence of the British Biological Warfare Programme," *Minerva* 40 (2002): 57–75; Kenneth V. Thimann, "The Role of Biologists in Warfare," *Bulletin of the Atomic Scientists* 3, 8 (Aug. 1947): 211–12; Jeanne Guillemin, "Scientists and the History of Biological Weapons," *European Molecular Biology Organization Reports* 7 (2006): S45–S49.

8. Peter Williams and David Wallace, *Unit 731: Japan's Secret Biological Warfare in World War II* (New York: The Free Press, 1989), 37–38.

9. Bruno Latour and Steve Woolgar, *Laboratory Life: the Construction of Scientific Facts* (Princeton, NJ: Princeton University Press, 1986), 200–202.

10. Gradon B. Carter and Graham S. Pearson, "British Biological Warfare and Biological Defence, 1925–45," in Geissler and Moon, *Biological and Toxin Weapons*, 168–89, see 177.

11. D. W. Henderson, "The Microbiological Research Department, Ministry of Supply, Porton, Wilts," *Proceedings of the Royal Society of London. Series B, Biological Sciences* 143, 911 (Jan. 27, 1955): 192–202, see 194.

12. Robert E. Kohler, *Lords of the Fly: Drosophila Genetics and the Experimental Life* (Chicago: University of Chicago Press, 1994), 6, 9–11.

13. Mark Wheelis, "First Shots Fired in Biological Warfare," correspondence, *Nature* 395 (Sept. 17, 1998): 213; Wheelis, "Biological Warfare before 1914," in Geissler and Moon, *Biological and Toxin Weapons*, 8–34; "Biological Sabotage in World War I," in Geissler and Moon, *Biological and Toxin Weapons*, 35–62. The group of scholars that contributed to the SIPRI series has done the best primary research on the BW

programs of several nations. I recommend their work as further reading and rely on it in this chapter.

14. Wheelis, "Biological Sabotage," 37.

15. W. E. Farrar, "Anthrax from Mesopotamia to Molecular Biology," *Pharos Alpha Omega* 58, 2 (1995): 35–38.

16. Wheelis, "First Shots."

17. Caroline Redmond, Martin J. Pearce, Richard J. Manchee, and Bjorn P. Berdal, "Deadly Relic of the Great War," *Nature* 393 (June 25, 1998): 747–48. The spores, although slow growers, tested positive for the ability to make both anthrax toxins.

18. Nadolny, trained as a lawyer, served during his career as a judge, foreign service officer, and German ambassador to Turkey and Moscow. During World War I, he served on the general staff.

19. Wheelis, "First Shots."

20. Wheelis describes a possible attempt to establish a German production facility for biological weapons in Spain but has found no evidence that this was actually realized. See "Biological Sabotage," 51.

21. Ibid., 40–41. Wheelis based his assessment on archival research, including an interview with Dilger found in Personal File 19368, July 29, 1917, Records of the WFGS, Military Intelligence Division Correspondence 1917–1941, Record Group 165, National Archives and Records Administration, Suitland, Maryland (hereafter NARA-S); and Records Relating to the Sabotage Claims Filed with the [Mixed Claims] Commission, Record Group 76, NARA-S. For a recent book-length account of Dilger's activities, see Robert Koenig, *The Fourth Horseman: The Tragedy of Anton Dilger and the Birth of Biological Terrorism* (New York: Public Affairs, 2007).

22. "Memorandum re record data re Carl Dilger with specific respect to the record as it existed at the time of the decision of October 16, 1930," Nov. 12, 1935, p. 4, in box 3, RG 76, entry 29 (Records Relating to the Sabotage Claims Filed with the [Mixed Claims] Commission), NARA-S.

23. Thomas Bonner, *Becoming a Physician: Medical Education in Britain, France, Germany, and the United States, 1750–1945* (New York: Oxford University Press, 1995). George W. Merck, later head of the American biological weapons development program during World War II, wished to attain his scientific training in Germany in 1915 but was prevented from doing so by the outbreak of World War I.

24. Anton Dilger's personal effects and passport were confiscated by the German government upon his death. German records were destroyed toward the end of the war, leaving us with incomplete information. Wheelis and other sources say Dilger was a victim of the influenza pandemic, but Chad Millman implies that Dilger might have been poisoned by German agents. Millman, *The Detonators: The Secret Plot to Destroy America and an Epic Hunt for Justice* (Boston, MA: Little, Brown, 2006), 165.

25. "Memorandum re record data re Carl Dilger"; "Examination of Paul G. L. Hilken," Aug. 26, 1930, exhibit 829 in entry 11 (U.S. Exhibits), RG 76, NARA-S; "Examination of Paul G. L. Hilken," Dec. 1928, exhibit 583 in entry 14 (printed copy of U.S. Exhibits), RG 76, NARA-S. All of these sources are also cited by Wheelis in "Biological Sabotage."

26. Wheelis, "Biological Sabotage," 41. Unlike Wheelis' interpretation, I think it is possible that Dilger's cultures did not perform as expected. The results of the inoculations in horses bound for France would not have been evident for several days afterward, and information about the outcome would have been available only with difficulty. Given what we know about the life cycle of *B. anthracis* and the reported conditions under which Dilger propagated it, his patriotic enthusiasm may have led him to exaggerate his success.

27. J. Edward Felton, exhibit 761, April 11, 1930, box 8, RG 76, Records of the Mixed Claims Commission, entry 11 (U.S. Exhibits), NARA-S.

28. Ibid.

29. Rosebury and Kabat, "Bacterial Warfare," 13.

30. Wheelis, "Biological Sabotage," 44–45.

31. George W. Merck, "Report to the Secretary of War on Biological Warfare (1943)," published after the war in *Military Surgeon* 98 (March 1946): 237–42, quotation on 237 (hereafter the Merck Report); reprinted in *Bulletin of the Atomic Scientists* 2 (Oct. 1, 1946): 16–22. The report was also excerpted in Sidney Shalett, "U.S. Was Prepared to Combat Axis in Poison-Germ Warfare," *New York Times*, Jan. 4, 1946, 13.

32. For more on the nonbiological sabotage campaign, see Henry Landau, *The Enemy Within: The Inside Story of German Sabotage in America* (New York: G. P. Putnam's Sons, 1937); Jules Witcover, *Sabotage at Black Tom: Imperial Germany's Secret War in America, 1914–1917* (Chapel Hill, N.C.: Algonquin Books, 1989); and Millman, *The Detonators.*

33. The American Veterinary Medical Association recorded no deliberations on the possibility of biological agents used against animals; nor did its major means of communication with American veterinarians, the *Journal of the American Veterinary Medical Association,* advise its readers to watch for unusual disease outbreaks among livestock.

34. Henry Field Smyth, "Anthrax—A Continuing and Probably Increasing Hazard of Industry," *American Journal of Public Health* 14 (November 1924): 920–24, see 921; "Anthrax in Man," *Annual Report of the Surgeon General of the Public Health Service of the United States* (Washington, D.C.: Government Printing Office, 1917): 259–60.

35. Wheelis, "Biological Sabotage," 51–54.

36. Kaiserlicher Geschäftsträger, Telegramm an Auswärtiges Amt [für Großer Generalstab Politik], Entzifferung, R 21241, Jan. 6, 1917, p. 36, Politisches Archiv des Auswärtigen Amts, Bonn, cited in Wheelis, "Biological Sabotage," 55.

37. Kleine became director in 1933. For the announcement in English, see *Science* 78, 2019 (Sept. 8, 1933): 208.

38. On Kleine, see F. K. Kleine, *Ein Deutscher Tropenarzt* [Einführung von Herbert Kunert] (Hannover: Schmorl and von Seefeld, 1949); M. Beck and F. K. Kleine: *Bericht über die Tätigkeit der zur Erforschung der Schlafkrankheit im Jahre 1906/07 nach Ostafrica entsandten Kommission* (Arbeiten aus dem Kaiserlichen Gesundheitsamte, 1909) 31: 1–319.

39. Christoph Gradmann, *Krankheit im Labor: Robert Koch und die medizinische Bakteriologie* (Göttingen: Wallstein Verlag, 2005).

40. Olivier Lepick, "French Activities Related to Biological Warfare," in Geissler and Moon, *Biological and Toxin Weapons,* 70–90, see 70, 74, 75.

41. Guillemin, *Biological Weapons,* 24–25.

42. Michel Macheboef, "M. A. Trillat (1861–1944)," *Annales de l'Institut Pasteur* 75 (1947): 622–23.

43. Guillemin, *Biological Weapons,* vii.

44. Lepick, "French Activities," 73.

45. Martha L. Hildreth, "The Influenza Epidemic of 1918–19 in France: Contemporary Concepts of Aetiology, Therapy, and Prevention," *Social History of Medicine* 4, 2 (Aug. 1991): 277–94, see 283.

46. Lepick, "French Activities," 77; Guillemin, *Biological Weapons,* 25.

47. Guillemin, *Biological Weapons,* 25–26.

48. Ibid., 26.

49. Costedoat had a special interest in military matters; he had authored articles on the subjects of bombs in cities and biological weapons. See for example A. Costedoat, "Les Bombardements Explosifs et Incendiaires des Villes," *La Presse Médicale* 58 (Aug. 19, 1950): 908–13; "Les Recherches Scientifiques sur la Protection Médicale de la Population contre la Guerre," ibid., 913–15.

50. Olivier Lepick, "The French Biological Weapons Program," *Deadly Cultures: Biological Weapons Since 1945,* ed. Mark Wheelis, Lajos Rózsa, and Malcolm Dando (Cambridge, MA: Harvard University Press, 2006), chapter 5, see 112–13.

51. H. Buchner and Friedrich Merkel, "Versuche über Inhalation trocken zerstäubter Milzbrandsporen," *Archiv für Hygiene* 8 (1888): 165–90; H. Buchner and E. Enderlen, "Inhalation von nass-zerstäubten Milzbrand-Sporen und—Stäbchen und von Hühnercholerabacillen," *Archiv für Hygiene* 8 (1888): 190–217. For the information on artificial anthrax pneumonia, see Buchner and Enderlen, 210.

52. Based on the research of Han Xiao, then deputy director of the Ping Fan 731 Museum, Sheldon H. Harris has also described the existence of a human experimentation center in Manchuria that predated Ping Fan: Beinyinhe Bacterial Factory, dating to 1932. See Harris, *Factories of Death: Japanese Biological Warfare 1932–45 and the American Cover-up* (London and New York: Routledge, 1994), 4, 30.

53. On the Tuskegee study, see James H. Jones, *Bad Blood: The Tuskegee Syphilis Experiment* (New York: Free Press, 1993); Susan M. Reverby, ed., *Tuskegee's Truths:*

Rethinking the Tuskegee Syphilis Study (Chapel Hill and London: University of North Carolina Press, 2000).

54. For an excellent analysis of views about human experimentation among the combatants at this time, see Gerhard Baader, Susan E. Lederer, Morris Low, Florian Schmaltz, Alexander V. Schwerin, "Pathways to Human Experimentation, 1933–1945: Germany, Japan, and the United States," *Osiris* 20 (2005): 205–31.

55. This is likely to be an underestimate; see Harris, *Factories of Death,* 66–67.

56. John W. Powell, "A Hidden Chapter in History," *Bulletin of the Atomic Scientists* 37, 8 (Oct. 1981): 44–52, see 47.

57. Robert Gomer, John W. Powell, Bert V. A. Röling, "Japan's Biological Weapons: 1930–1945," *Bulletin of the Atomic Scientists* 37, 8 (Oct. 1981): 43; Powell, "A Hidden Chapter," 46.

58. Powell, "A Hidden Chapter," 44.

59. Gomer, Powell, and Röling, "Japan's Biological Weapons."

60. Bert V. A. Röling, "A Judge's View," *Bulletin of the Atomic Scientists* 37, 8 (Oct. 1981): 52–53, quotation on 53.

61. American scientists brought over to interview Japanese BW officials pointed out that they had spent very little money in the process of acquiring "information [that] could not be obtained in our own laboratories because of scruples attached to human experimentation"; in Powell's words, they viewed the acquisition of the data through the immunity deal as a "bargain." Powell, "A Hidden Chapter," 46–47. Despite initial excitement, the data were later judged to be disappointing.

62. Powell, "A Hidden Chapter," 51.

63. See, for example, Harris, *Factories of Death,* 64, 66.

64. Ibid., 69.

65. Ibid., 68–69.

66. Sheldon Harris, "The Japanese Biological Warfare Program," in Geissler and Moon, *Biological and Toxin Weapons,* 127–52, see 139.

67. Harris, *Factories of Death,* 61.

68. Kei-ichi Tsuneishi, *The Biological Warfare Unit that Disappeared* (Kieta Saikinsen Butai) (Tokyo: Kai-mei-sha Publishers, 1981), 132–33. Tsuneishi and Seii-chi Morimura have been the leading Japanese scholars analyzing biological weapons tests in Manchuria. For more on the Japanese historiography and political context, see Frederick R. Dickinson, "Biohazard: Unit 731 in Postwar Japanese Politics of National 'Forgetfulness,'" in *Dark Medicine: Rationalizing Unethical Medical Research,* ed. William R. LaFleur, Gernot Böhme, and Susumu Shimazono (Bloomington and Indianapolis: Indiana University Press, 2007), 85–104.

69. Williams and Wallace, *Unit 731,* 5–8.

70. Ibid., 94. Fox's skepticism did not dampen Ishii's enthusiasm.

71. Harris, "The Japanese Biological Warfare Program," 137.

72. Akimoto quoted in Williams and Wallace, *Unit 731,* 38. During the 1920s, several nations legalized or considered offering criminal prisoners the opportunity to

participate in medical experimentation in return for reduction of their sentences. See Susan E. Lederer, *Subjected to Science: Human Experimentation in America before the Second World War* (Baltimore, MD: Johns Hopkins University Press, 1995), 111–13. This does not in any way excuse Japanese activities in Manchuko, but it does remind us that early twentieth-century ethical guidelines differed greatly from those we recognize today, a century later.

73. Ryōhei Sakaki, "Bacteriological Warfare," *Sunday Mainichi*, no. 1682, Jan. 27, 1952, reproduced in Williams and Wallace, *Unit 731*, 37–38.

74. Williams and Wallace, *Unit 731*, 91–93.

75. Ibid., 92.

76. Rosebury, *Peace or Pestilence*, 68–69.

77. Lepick, "French Activities," 84–85.

78. Martin Hugh-Jones, "Wickham Steed and German Biological Warfare Research," *Intelligence and National Security* 7 (1992): 379–402, see 395; Carter and Pearson, "British Biological Warfare," 170–73.

79. Carter and Pearson, "British Biological Warfare," 176.

80. Brian Balmer, "The UK Biological Weapons Program," in Wheelis et al., *Deadly Cultures*, 47–83, see 48.

81. John Ellis van Courtland Moon, "U.S. Biological Warfare Planning and Preparedness: the Dilemmas of Policy," in in Geissler and Moon, *Biological and Toxin Weapons*, 215–54, see 219.

82. For particulars on the advantages to each of the trilateral partners, see Donald Avery, "Canadian Biological and Toxin Warfare Research, Development, and Planning, 1925–45," in Geissler and Moon, *Biological and Toxin Weapons*, 190–214; also chapters 9 and 11 in the same volume.

83. Avery, "Canadian Biological and Toxin Warfare Research," 201–202. Hagan worried about how much information should be conveyed to veterinarians, whom he viewed as a frontline defense against sabotage attacks on livestock. See Ellis Pierson Leonard, *In the James Law Tradition 1908–1948* (Ithaca, NY: Cornell University Press, 1982), 324–25.

84. Notes, *Veterinary Medicine* 42 (March 1947): 90.

85. L. H. Kent and W. T. J. Morgan, "David Willis Wilson Henderson, 1903–1968," *Biographical Memoirs of Fellows of the Royal Society*, 16 (Nov. 1970): 331–41.

86. This was easier said than done. Different species and strains of bacilli were notoriously difficult to separate from each other, a topic that considered in more depth in chapter 5. See C. S. Pederson, "Symposium on Problems in Taxonomy," *Bacteriological Reviews* 20 (1956): 274–76.

87. Sidney Auerbach and George G. Wright, "Studies on Immunity in Anthrax. VI. Immunizing Activity of Protective Antigen Against Various Strains of *Bacillus Anthracis*," *Journal of Immunology* 75, 2 (1955): 129–33, see 129.

88. Geoffrey Holland, "United States Exports of Biological Materials to Iraq: Compromising the Credibility of International Law," e-published, with citations, at

http://deepblade.net/journal/Holland_JUNE2005.pdf, June 2005, 13–15. Martin Hugh-Jones recollected the strain's history.

89. John Ellis van Courtland Moon, "U.S. Biological Warfare Planning and Preparedness: the Dilemmas of Policy," *Biological Weapons*, 215–54, see 245.

90. H. A. Druett, D. W. Henderson, L. Packman, and S. Peacock, "Studies on Respiratory Infection I. The Influence of Particle Size on Respiratory Infection with Anthrax Spores," *Journal of Hygiene* 51 (Sept. 1953): 359–71, see 360. Henderson and his colleague C. E. Venzke had isolated M-36 in 1944 but, not surprisingly, did not publish the information.

91. Pederson, "Symposium on Problems in Taxonomy."

92. S. Auerbach and G. G. Wright, "Studies on Immunity in Anthrax. VI. Immunizing Activity of Protective Antigen against Various Strains of *Bacillus anthracis*," *Journal of Immunology* 75 (1955): 129–33; Stephen F. Little and Gregory B. Knudsen, "Comparative Efficacy of *Bacillus anthracis* Live Spore Vaccine and Protective Antigen Vaccine Against Anthrax in the Guinea Pig," *Infection and Immunity* 52 (May 1986): 509–12.

93. Ken Alibek, with Stephen Handelman, *Biohazard: The Chilling True Story of the Largest Covert Biological Weapons Program in the World—Told from Inside by the Man who Ran It* (New York: Dell Publishing, 1999), 132.

94. Scott Shane, "Army Harvested Victims' Blood to Boost Anthrax: Ex-scientists Detail Detrick Experiments," *Baltimore Sun*, Dec. 23, 2001, 1A. The two deaths are not disclosed in open source materials; the Merck Report and Rosebury's *Peace or Pestilence,* which reviewed the technical literature, mention twenty-five cases of cutaneous anthrax and no deaths. See Rosebury, *Peace or Pestilence,* 49n.

95. Rosebury, *Peace or Pestilence,* 48.

96. Paul Keim, Abdulahi Kalif, James Schupp, Karen Hill, Steven E. Travis, Kara Richmond, Debra M. Adair, Martin Hugh-Jones, Cheryl R. Kuska, and Paul Jackson. "Molecular Evolution and Diversity in *Bacillus anthracis* as Detected by Amplified Fragment Length Polymorphism Markers," *Journal of Bacteriology* 179 (1997): 818–24; Lyndsay Radnedge, Peter G. Agron, Karen K. Hill, Paul J. Jackson, Lawrence O. Ticknor, Paul Keim, and Gary L. Andersen, "Genome Differences that Distinguish *Bacillus anthracis* from *Bacillus cereus* and *Bacillus thuringiensis,*" *Applied and Environmental Microbiology* 69, 5 (May 2003): 2755–64.

97. Theodor Rosebury, *Experimental Air-Borne Infection* (Baltimore, MD: William and Wilkins, 1947), v.

98. See for example J. M. Barnes, "The Development of Anthrax Following the Administration of Spores by Inhalation," *British Journal of Experimental Pathology* 28 (1947): 385–94; Gradon B. Carter and Graham S. Pearson, "British Biological Warfare and Biological Defence, 1925–45," in Geissler and Moon, *Biological and Toxin Weapons,* 168–89, see 176–77.

99. D. W. Henderson, "An Experimental Apparatus for the Study of Airborne Infection," *Journal of Epidemiology and Infection* 50, 1 (March 1952): 53–68, quota-

tion on 53. For the "piccolo" designation, see Kent and Morgan, "David Willis Wilson Henderson," 335. This nickname probably referred to the animal exposure tube, which was a metal tube about 26 inches long that resembled the musical instrument.

100. Druett et al., "Studies on Respiratory Infection I," 359.

101. Ibid., 360.

102. Ibid., 359.

103. Henderson, "An Experimental Apparatus," 58–59.

104. My speculations, based on work done in experimental exercise physiology in the Harvard Concord Field Station laboratory of the late C. Richard Taylor.

105. Henderson, "An Experimental Apparatus," 67.

106. Ibid., 55.

107. Ibid., 54–55.

108. Ibid., 55–56.

109. Rosebury, *Experimental Air-Borne Infection,* 60–66; Henderson, "An Experimental Apparatus," 62.

110. Henderson, "An Experimental Apparatus," 67.

111. Druett et al., "Studies on Respiratory Infection I," 360; Barnes, "The Development of Anthrax," 385.

112. Druett et al., "Studies on Respiratory Infection I," 370.

113. According to Fort Detrick Lieutenant Colonel Murray Sanders, interviewed in Williams and Wallace, *Unit 731,* 123; see also Carter and Pearson, "British Biological Warfare," 177n40.

114. Rosebury, *Experimental Air-Borne Infection,* v.

115. Williams and Wallace, *Unit 731,* 123; Kent and Morgan, "David Willis Wilson Henderson," 339.

116. Avery, "Canadian Biological and Toxin Warfare," 202.

117. Ibid., 203.

118. Moon, "U.S. Biological Warfare Planning," 237.

119. Avery, "Canadian Biological and Toxin Warfare," 205–207.

120. Daniel H. Fine, "Dr. Theodor Rosebury: Grandfather of Modern Oral Microbiology," *Journal of Dental Research* 85, 11 (2006): 990–95. Rosebury was a dentist who had completed a fellowship in biological chemistry at Columbia and then joined the bacteriology department of the College of Physicians and Surgeons there.

121. Rosebury and Kabat, "Bacterial Warfare," 11.

122. Ibid., 9.

123. Rosebury, *Experimental Air-Borne Infection,* 4–5.

124. Ibid., v, 57, 92.

125. Rosebury, *Peace or Pestilence,* 68.

126. Moon, "U.S. Biological Warfare Planning," 245. Moon cites an unpublished manuscript: R. C. Cochrane, "History of the Chemical Warfare Service in

World War II," vol. 2, "Biological Warfare Research in the United States" (Historical Section: Plans, Training, and Intelligence Division, Office of the Chief, Chemical Corps, Nov. 1947), secured from the Historical Office of Aberdeen Proving Ground, Aberdeen, MD.

127. Williams and Wallace, *Unit 731*, 123.

128. Rosebury, *Peace or Pestilence*, 195.

129. Ibid., 195–96.

130. Moon, "U.S. Biological Warfare Planning," 251.

131. Wheelis, "Biological Sabotage," 37–38.

132. This point differs from the analysis of Guillemin, who states that the microbial cloud paradigm "quickly replaced" the older delivery systems based on chemical weapons (see Guillemin, *Biological Weapons*, vii).

133. Rosebury and Kabat, "Bacterial Warfare," 12.

134. Rosebury, *Peace or Pestilence*, 36.

135. Balmer, "The UK Biological Weapons Program," 48.

136. Paul Keim and K. L. Smith, "*Bacillus anthracis* Evolution and Epidemiology," in *Anthrax*, ed. Theresa M. Koehler (Berlin: Springer-Verlag, 2002), 21–31, see 21–22.

137. Julian Parkhill and Colin Berry, "Relative Pathogenic Values," News and Views section, *Nature* 423 (May 1, 2003): 23–25, quotation on 25.

138. Thanks to Christine Manganaro for helpful discussions on these points.

139. Guillemin, *Biological Weapons*, viii

140. Rosebury, *Peace or Pestilence*, 61.

Chapter 5. Resistance

Epigraph: Richard M. Nixon, "Remarks Announcing Decisions on Chemical and Biological Defense Policies and Programs," Nov. 25, 1969.

1. Lynne Roberts, "The Deadly Legacy of Anthrax Island," *London Times Sunday Magazine*, Feb. 15, 1981, 20–27, quotation on 20.

2. Notes, *Veterinary Medicine* 42 (March 1947): 90.

3. Ferenc M. Szasz, "The Impact of World War II on the Land: Gruinard Island, Scotland, and Trinity Site, New Mexico, as Case Studies," *Environmental History Review* 19, 4 (Winter 1995): 15–30, see 20–21. Along with Gruinard Island, *B. anthracis* bomblets were tested on a beach at Penclawdd. See Peter M. Hammond and Gradon Carter, *From Biological Warfare to Healthcare: Porton Down, 1940–2000* (Basingstoke, UK: Palgrave, 2002), 63.

4. Hammond and Carter, *From Biological Warfare to Healthcare*, 10.

5. Brian Balmer, "The UK Biological Weapons Program," *Deadly Cultures: Biological Weapons Since 1945*, ed. Mark Wheelis, Lajos Rózsa, and Malcolm Dando (Cambridge, MA: Harvard University Press, 2006), 47–83, see 56–60.

6. Hammond and Carter, *From Biological Warfare to Healthcare*, 40.

7. Bruno Latour, *The Pasteurization of France*, trans. Alan Sheridan and John Law (Cambridge, MA: Harvard University Press, 1988), 77 and 85–87.

8. Richard M. Nixon, "Remarks Announcing Decisions on Chemical and Biological Defense Policies and Programs," Nov. 25, 1969; Interdepartmental Political-Military Group, "Report to the National Security Council: U.S. Policy on Chemical and Biological Warfare and Agents," Nov. 10, 1969 (declassified on April 25, 1996), 24–28, www.gwu.edu/%7Ensarchiv/NSAEBB/NSAEBB58/RNCBW6a.pdf (accessed Aug. 20, 2008). At this time, the United States was spending $36.4 million annually on biological programs. See also Amy E. Smithson, "Biological Weapons: Can Fear Overwhelm Inaction?" *Washington Quarterly* 28, 1 (2004–2005): 165–78.

9. Jeanne Guillemin, *Anthrax: The Investigation of a Deadly Outbreak* (Berkeley: University of California Press, 1999).

10. Statistics summarized in Hammond and Carter, *From Biological Warfare to Healthcare*, 6.

11. Guillemin, *Anthrax*, 243.

12. U.S. Congress, Office of Technology Assessment, *Technologies Underlying Weapons of Mass Destruction, OTA-BP-ISC-115* (Washington, D.C.: U.S. Government Printing Office, Dec. 1993), 70.

13. The best overview is Mark Wheelis, Lajos Rósza, and Malcolm Dando, eds., *Deadly Cultures: Biological Weapons Since 1945* (Cambridge, MA: Harvard University Press, 2006). See page 2 for a summary of postwar BW programs.

14. Theodor Rosebury, *Peace or Pestilence: Biological Warfare and How to Avoid It* (New York: McGraw-Hill, 1949), 36.

15. William M. Creasy, "Biological Warfare," *Armed Forces Chemical Journal* 5 (Jan. 1952): 16–18, 46, see 18.

16. See, for example, U.S. Congress, Office of Technology Assessment, *Technologies Underlying Weapons of Mass Destruction*, 70.

17. Robert E. Kohler, *Lords of the Fly:* Drosophila *Genetics and the Experimental Life* (Chicago: University of Chicago Press, 1994).

18. An idea dating back to at least Plato, second nature has gained modern prominence through the ideas of Karl Marx and Friedrich Engels. For more on this concept, see Aletta Biersack and James B. Greenberg, eds., *Reimagining Political Ecology* (Durham, NC: Duke University Press, 2006), 13–14; Reiner Grundmann, "The Ecological Challenge to Marxism," *New Left Review* 187 (May–June 1991): 103–20; William Cronon, *Nature's Metropolis: Chicago and the Great West* (New York: W.W. Norton, 1991), see esp. xix.

19. Hannah Landecker, *Culturing Life: How Cells Became Technologies* (Cambridge, MA: Harvard University Press, 2007).

20. Rosebury, *Peace or Pestilence*, 36.

21. Theodor Rosebury and Elvin A. Kabat, "Bacterial Warfare: A Critical Analysis of the Available Agents, Their Possible Military Applications, and the Means for Protection Against Them," *Journal of Immunology* 56, 1 (May 1947): 7–96.

22. Rosebury, *Peace or Pestilence*, 37–38.

23. See the Merck Report, 1946, which mentioned around sixty cases. By 1949,

Rosebury (*Peace or Pestilence,* 49) tallied another fifty-six cases of accidental disease among laboratory personnel, twenty-five of which were anthrax infections. George W. Merck, "Report to the Secretary of War on Biological Warfare (1943)," *Military Surgeon* 98 (March 1946): 16–18; *Bulletin of the Atomic Scientists* 2 (Oct. 1, 1946): 16–18; condensed in the *New York Times,* Jan. 4, 1946, 13.

24. Rosebury, *Peace or Pestilence,* 38.

25. Valentin Bojtzov and Erhard Geissler, "Military Biology in the USSR," *Biological and Toxin Weapons: Research, Development and Use from the Middle Ages to 1945,* ed. Geissler and John Ellis van Courtland Moon, SIPRI (Stockholm International Peace Research Institute) Chemical and Biological Warfare Studies No. 18 (Oxford: Oxford University Press, 1999), 153–67, see 165–66.

26. Balmer, "The UK Biological Weapons Program," 47.

27. U.S. Congress, Office of Technology Assessment, *Technologies Underlying Weapons of Mass Destruction,* 72.

28. Millet, "Anti-Animal Biological Weapons Programs," 233.

29. This point is forcefully made by Leonard A. Cole, *The Eleventh Plague: The Politics of Biological and Chemical Warfare* (New York: Freeman, 1997), 199–202.

30. Henderson quoted in Balmer, "The UK Biological Weapons Program," 48.

31. Ibid., 54.

32. Cited in Ibid., 50.

33. D. W. Henderson, "The Microbiological Research Department, Ministry of Supply, Porton, Wilts," *Proceedings of the Royal Society of London, Biological Sciences,* 143, 911 (Jan. 27, 1955): 192–202, quotation on 202.

34. See Balmer, "The UK Biological Weapons Program," 56–60, for descriptions of the field tests.

35. Ibid., 61–63.

36. Ibid., 67.

37. Henderson, "The Microbiological Research Department," 192.

38. Ibid., 193.

39. Ibid., 196.

40. J[oan] M. Ross, "On the Histopathology of Experimental Anthrax in the Guinea-Pig," *British Journal of Experimental Pathology* 36, 3 (June 1955): 336–39; D. W. Henderson, S. Peacock, F. C. Belton, "Observations on the Prophylaxis of Experimental Pulmonary Anthrax in the Monkey," *Journal of Hygiene,* 54, 1 (March 1956): 28–36.

41. Henderson, "The Microbiological Research Department," 196.

42. O[tto] Bollinger, "Infectionen durch thierische Gifte: Milzbrand," *Handbuch der Chronischen Infectionskrankheiten* (Leipzig: Verlag von F.C. W. Vogel, 1874); Bollinger, "Anthrax," in H. von Ziemssen, A. H. Buck, and G. L. Peabody, *Cyclopedia of the Practice of Medicine* (New York: William Wood, 1874–1881), vol. 3 (1875), 372–430, see 374.

43. Harry Smith, "Discovery of the Anthrax Toxin: The Beginning of Studies of

Virulence Determinants Regulated *in Vivo*," *International Journal of Medical Microbiology* 291 (2002): 411–17.

44. H. Smith, J. Keppie, and J. L. Stanley, "The Chemical Basis of the Virulence of *Bacillus anthracis*. V. The Specific Toxin Produced by *B. anthracis* in Vivo," *British Journal of Experimental Pathology* 36 (1955): 460–72; P. W. Harris-Smith and J. Keppie, "Production *in Vitro* of the Toxin of *Bacillus anthracis* Previously Recognized *in Vivo*," *Journal of General Microbiology* 19 (1958): 91–103; Henderson, "The Microbiological Research Department," 196.

45. G. P. Gladstone, "Immunity to Anthrax: Protective Antigen Present in Cell-Free Culture Filtrates," *British Journal of Experimental Pathology* 27 (1946): 394–418.

46. F. C. Belton and R. E. Strange, "Studies on a Protective Antigen Produced *in Vitro* from *Bacillus anthracis:* Medium and Method of Production," *British Journal of Experimental Pathology* 35 (1954): 144–52.

47. Peter C. B. Turnbull, "Anthrax Vaccines: Past, Present, and Future," *Vaccine* 9 (Aug. 1991): 533–39, see 534; Turnbull, "Introduction: Anthrax History, Disease and Ecology," in *Anthrax,* ed. T. M. Koehler (Berlin: Springer-Verlag, 2002), 1–19, see 4.

48. Henderson, "The Microbiological Research Department," 197.

49. Rosebury, *Peace or Pestilence,* 185. Rosebury sharply distinguished the "wartime emergency" situation from that of the cold war.

50. Daniel H. Fine, "Dr. Theodor Rosebury: Grandfather of Modern Oral Microbiology," *Journal of Dental Research* 85, 11 (2006): 990–95.

51. John Ellis van Courtland Moon, "The U.S. Biological Weapons Program," in Wheelis et al., *Deadly Cultures,* 9–46, see 19.

52. Stephen Endicott and Edward Hagerman, *The United States and Biological Warfare: Secrets from the Early Cold War and Korea* (Bloomington: Indiana University Press, 1998), 195.

53. Moon, "The U.S. Biological Weapons Program," 24.

54. Judith Miller, Stephen Engelberg, and William Broad, *Germs: Biological Weapons and America's Secret War* (New York: Simon and Schuster, 2001), 44–45. For an account of the Australian rabbit destruction and the virus that caused it, see F. J. Fenner and F. N. Ratcliffe, *Myxomatosis* (Cambridge: Cambridge University Press, 1965).

55. "New Anthrax Vaccine," *Science News Letter* 67 (June 4, 1955): 354.

56. G. G. Wright and J. B. Slein, "Studies on Immunity in Anthrax. I. Variation in the Serum T-Agglutinin during Anthrax Infection in the Rabbit," *Journal of Experimental Medicine* 93 (1951): 99–106; G. G. Wright, M. A. Hedberg, and R. J. Feinberg, "Studies on Immunity in Anthrax. II. *In Vitro* Elaboration of Protective Antigen by Nonproteolytic Mutants of *Bacillus anthracis,*" *Journal of Experimental Medicine* 93 (1951): 523–27; G. G. Wright, M. A. Hedberg, and J. B. Slein, "Studies on Immunity in Anthrax. III. Elaboration of Protective Antigen in a Chemically-Defined Nonprotein Medium," *Journal of Immunology* 72 (1954): 263–69.

57. Philip S. Brachman, H. Gold, S. A. Plotkin, F. R. Rekety, M. Werrin, and N. R. Ingraham, "Field Evaluation of a Human Anthrax Vaccine," *American Journal of Public Health* 52 (1962): 632–45.

58. R. E. Strange and C. B. Thorne, "Further Purification Studies on the Protective Antigen of *Bacillus anthracis* Produced in Vitro," *Journal of Bacteriology* 76 (1958): 192–202, see 192.

59. J. L. Stanley and Harry Smith, "Purification of Factor I and Recognition of a Third Factor of the Anthrax Toxin," *Journal of General Microbiology* 26 (1961): 49–66; F. A. Beall, Martha J. Taylor, and C. B. Thorne, "Rapid Lethal Effect in Rats of a Third Component Found upon Fractionating the Toxin of *Bacillus anthracis*," *Journal of Bacteriology* 83 (1962): 1274–80; Harry Smith and H. B. Stoner, "Anthrax Toxic Complex," *Federation Proceedings* 26 (1967): 1554–57; F. Klein, I. A. DeArmon, Jr., R. E. Lincoln, B. G. Mahlandt, A. L. Fernlius, "Immunological Studies of Anthrax. II. Levels of Immunity against *Bacillus anthracis* Obtained with Protective Antigen and Live Vaccine," *Journal of Immunology* 88 (1962): 15–19.

60. Landecker, *Culturing Life: How Cells Became Technologies,* 16.

61. Togzhan Kassenova, "Biological Threat Reduction in Central Asia," *Bulletin of the Atomic Scientists,* www.thebulletin.org/node/3711 (accessed Aug. 14, 2008).

62. V. P. Shuylak, "[Epidemiological Efficacy of Anthrax STI Vaccine in Tadjik SSR]," *Zhurnal mikrobiologii, epidemiologii i immunobiologii* 47 (1970): 117–20 (in Russian); E. Shlyakhov and E. Rubenstein, "Human Live Anthrax Vaccine in the Former USSR," *Vaccine* 12 (1994): 727–30.

63. Guillemin, *Biological Weapons,* 136.

64. Ken Alibek, with Stephen Handelman, *Biohazard: The Chilling True Story of the Largest Covert Biological Weapons Program in the World—Told from Inside by the Man who Ran It* (New York: Dell, 1999), 160.

65. John Hart, "The Soviet Biological Weapons Program," in Wheelis et al., *Deadly Cultures,* 132–56, see 143–44, 154.

66. See, for example, B. C. J. Knight and H. Proom, "A Comparative Survey of the Nutrition and Physiology of Mesophilic Species in the Genus *Bacillus*," *Journal of General Microbiology* 4, 3 (1950): 508–38.

67. J. Tomcsik, "Induced Mutation of *Bacillus anthracis*," presented at the Society for General Microbiology Meeting, 1950; published in *Journal of General Biology* 4 (1958): xvi–xvii.

68. N. R. Smith, T. Gibson, Ruth E. Gordon, and P. H. A. Sneath, "Type Cultures and Proposed Neotype Cultures of Some Species in the Genus *Bacillus*," *Journal of General Microbiology* 34 (1964): 269–72, quotation on 269.

69. See for example Erlendur Helgason, Ole Andreas Økstad, Dominque A. Caugant, Henning A. Johansen, Agnes Fouet, Michéle Mock, Ida Hegna, and Anne-Brit Kolsø, "*Bacillus anthracis, Bacillus cereus,* and *Bacillus thuringiensis*—One Species on the Basis of Genetic Evidence," *Applied Environmental Microbiology* 66 (2000): 2627–30.

70. Smith et al., "Type Cultures," 271; see for example Lyndsay Radnedge, Peter G. Agron, Karen K. Hill, Paul J. Jackson, Lawrence O. Ticknor, Paul Keim, and Gary L. Andersen, "Genome Differences that Distinguish *Bacillus anthracis* from *Bacillus cereus* and *Bacillus thuringiensis,*" *Applied and Environmental Microbiology* 69 (May 2003): 2755–64.

71. Karen Rader develops this argument on the epistemological choices of scientists developing and using standardized laboratory organisms in *Making Mice: Standardizing Animals for American Biomedical Research, 1900–1955* (Princeton, NJ: Princeton University Press, 2004).

72. B. K. Nordberg and W. Thorsell, "The Effect of Certain Enzyme Systems on the Capsule of *Bacillus anthracis,*" *Journal of Bacteriology* 69, 4 (1955): 367–71.

73. H. Smith, "Discovery of the Anthrax Toxin."

74. Perry Mikesell, Bruce E. Ivins, Joseph D. Ristroph, and Thomas M. Dreier, "Evidence for Plasma-Mediated Toxin Production in *Bacillus Anthracis,*" *Infection and Immunity* 39, 1 (Jan. 1983): 371–76.

75. Hugh B. Cole, John W. Ezzell, Jr., Kenneth F. Keller, and Ronald J. Doyle, "Differentiation of *Bacillus anthracis* and Other *Bacillus* Species by Lectins," *Journal of Clinical Microbiology* 19 (Jan. 1984): 48–53.

76. Alibek, *Biohazard,* 234.

77. G. B. Carter, *Chemical and Biological Defence at Porton Down, 1916–2000* (London: HMSO, 2000), 147.

78. Jim A. Davis and Anna Johnson-Winegar, "The Anthrax Terror: DOD's Number-One Biological Threat," in *The Gathering Biological Warfare Storm,* ed. Jim A. Davis and Barry R. Schneider (Westport, CT: Praeger, 2004), 49–64, see 57–58.

79. U.S. Congress, Office of Technology Assessment, *Technologies Underlying Weapons of Mass Destruction,* 70.

80. Guillemin, *Biological Weapons,* 151.

81. Alibek, *Biohazard,* 273–75.

82. "Some Tea, then Chemical Sally's Composure Slips," *Daily Telegraph,* March 29, 2003, www.telegraph.co.uk/news/worldnews/middleeast/iraq/1426116/Some-tea, -then-Chemical-Sally%27s-composure-slips.html (accessed Aug. 23, 2008); Toby Harnden, "'Mrs. Anthrax' Held by U.S. Forces," *Daily Telegraph,* May 5, 2003, www.telegraph.co.uk/news/worldnews/middleeast/iraq/1429285/Mrs-Anthrax-held-by-US-forces.html (accessed Aug. 23, 2008); Oliver Poole, "Saddam's Dr. Germ and Mrs. Anthrax to Be Released," *Daily Telegraph,* Dec. 20, 2005, www.telegraph.co.uk/news/worldnews/middleeast/iraq/1505948/Saddams-Dr-Germ-and-Mrs-Anthrax-to-be-released.html (accessed Aug. 23, 2008).

83. Keith Bradsher, "Senator Says U.S. Let Iraq Get Lethal Viruses," *New York Times,* Feb. 10, 1994, A9; Kevin Merida and John Mintz, "Rockville Firm Shipped Germ Agents to Iraq, Riegle Says," *Washington Post,* Feb. 10, 1994, A8; "Conflict Alleged for Head of Study on Gulf War Illness," *Baltimore Sun,* Nov. 29, 1996, 20A; Michael White, "UK Anthrax Strains 'Sold to Iraq,'" *Guardian,* April 3, 1998, 10;

William Blum, "Anthrax for Export: U.S. Companies Sold Iraq the Ingredients for a Witch's Brew," *Progressive*, April 1998, 18; Guillemin, *Biological Weapons*, 153–54.

84. Poole, "Saddam's Dr. Germ."

85. Guillemin, *Biological Weapons*, 155–56.

86. Tom Mangold and Jeff Goldberg, *Plague Wars: A True Story of Biological Warfare* (New York: St Martin's Press, 1999), 241–42; Tom Fennell, "The Grisly Realm of 'Dr. Death,'" *Maclean's* 112 (Oct. 18, 1999): 48–49. See the provocative account by journalists and filmmakers Bob Coen and Eric Nadler, *Dead Silence: Fear and Terror on the Anthrax Trail* (Berkeley, CA: Counterpoint, Transformer Films, 2009), 143–59, for more information and speculation about the South African biological weapons program.

87. "Biological Facilities: Onderstepoort Veterinary Institute," Feb. 2004, James Martin Center for Nonproliferation Studies, presented by NTI (Nuclear Threat Initiative) www.nti.org/e_research/profiles/SAfrica/Biological/2432_3539.html (accessed Aug. 23, 2008).

88. Mangold and Goldberg, *Plague Wars*, 257–58.

89. John Schofield, "The Shadow of 'Dr. Death,'" *Maclean's* 111 (July 13, 1998): 40; Fennell, "The Grisly Realm of 'Dr. Death.'"

90. Miller, Engelberg, and Broad, *Germs*, 150.

91. Max Sterne, "Distribution and Economic Importance of Anthrax," *Federation Proceedings* 26 (1967): 1493–95.

92. Statistics summarized in Peter Hammond and Gradon Carter, *From Biological Warfare to Healthcare: Porton Down, 1940–2000* (Basingstoke, UK: Palgrave, 2002), 6.

93. J. C. A. Davies, "A Major Epidemic of Anthrax in Zimbabwe, Part I," *Central African Journal of Medicine* 28 (1982): 291–98; Meryl Nass, "Anthrax Epizootic in Zimbabwe, 1978–80: Due to Deliberate Spread?" www.anthraxvaccine.org/zimbabwe.html (accessed Oct. 31, 2007); World Health Organization, The World Anthrax Data Site, www.vetmed.lsu.edu/whocc/ (accessed Oct. 16, 2007). Anthrax is hyperendemic in three locations on the African continent: the west coast (from Sierra Leone to Ghana); central (Niger, Chad, and Ethiopia); and south (Zambia and Zimbabwe).

94. Nass, "Anthrax Epizootic in Zimbabwe," 2 of 11.

95. J. C. A. Davies, "A Major Epidemic of Anthrax in Zimbabwe, Part II," *Central African Journal of Medicine* 29 (1983): 8–12

96. Nass, "Anthrax Epizootic in Zimbabwe," 6.

97. Ibid., 7.

98. Ibid., 9. Anthrax experts, including Mark Wheelis and Philip S. Brachman, read and commented on this article; Bruce E. Ivins also supplied Nass with information. Thus, the scientific community is well aware of Nass's theory.

99. Piers Millet, "Anti-Animal Biological Weapons Programs," 224–35 in Wheelis et al., *Deadly Cultures*, see 234–35.

100. Caroline Redmond, Martin J. Pearce, Richard J. Manchee, and Bjorn P. Berdal, "Deadly Relic of the Great War," *Nature* 393 (June 25, 1998): 747–48. The spores, although slow growers, tested positive for the ability to make both deadly anthrax toxins.

101. Hammond and Carter, *From Biological Warfare to Healthcare*, 11.

102. Shawn Pogatchnik, "Tourist Temptation: Anthrax Island," CBS News broadcast, Oct. 29, 2001.

103. Hammond and Carter, *From Biological Warfare to Healthcare*, 14.

104. Szasz, "The Impact of World War II on the Land," 21.

105. Hammond and Carter, *From Biological Warfare to Healthcare*, 13.

106. R. J. Manchee, M. G. Broster, J. Melling, R. M. Henstridge, and A. J. Stagg, "*Bacillus anthracis* on Gruinard Island," *Nature* 294 (Nov. 19, 1981): 254–45.

107. "Dark Harvest," *Time Magazine* 118, 19 (Nov. 9, 1981): 55–56, quotations on 56.

108. Szasz, "The Impact of World War II on the Land," 21.

109. R. J. Manchee, M. G. Broster, I. S. Anderson, and R. M. Henstridge, "Decontamination of *Bacillus anthracis* on Gruinard Island?" Letters, *Nature* 303 (May 19, 1983): 239–40.

110. "Britain's Anthrax Island," BBC News broadcast, July 25, 2001.

111. Hammond and Carter, *From Biological Warfare to Healthcare*, 15.

112. Szasz, "The Impact of World War II on the Land," 22.

113. Ibid., 27.

114. Christopher Pala, "Anthrax Island," *New York Times Magazine*, Jan. 12, 2003, 36–39.

115. GlobalSecurity.org, "Vozrozhdeniye Island, Rebirth/Renaissance Island," www.globalsecurity.org/wmd/world/russia/vozrozhdenly.htm (accessed Aug. 14, 2008).

116. NTI (Nuclear Threat Initiative), "Biological Facilities: Vozrozhdeniye Island, Aral Sea," www.nti.org/e_research/profiles/Kazakhstan/Biological/facilities_vozrozhdenie.html, (accessed Aug. 18, 2008).

117. GlobalSecurity.org, "Vozrozhdeniye Island, Rebirth/Renaissance Island."

118. Ibid.

119. Ken Alibek, May 23, 2000, testimony, Special Oversight Panel on Terrorism, House Armed Service Committee, 106th Congress, 2nd session; also cited in Eileen Choffnes, "Germs on the Loose: Bioweapons Tests Tainted Sites Around the World," *Bulletin of the Atomic Scientists* 57, 2 (March–April 2001): 57–61, see 61.

120. U.S. Department of Defense, *Cooperative Threat Reduction Program Annual Report to Congress, Fiscal Year 2009*, appendix B, available at www.dtra.mil/documents/oe/ctr/FY09%20CTR%20Annual%20Report%20to%20Congress.pdf (accessed Aug. 18, 2008); *The Biological Threat Reduction Program of the Department of Defense: From Foreign Assistance to Sustainable Partnerships* (Washington, D.C.: National Academies Press, 2007), 27–28, also cited in Togzhan Kassenova, "Biological Threat Reduction in Central Asia."

121. Kassenova, "Biological Threat Reduction in Central Asia."

122. U.S. Department of Defense, *Cooperative Threat Reduction Program Annual Report,* 16–20; Togzhan Kassenova, "Biological Threat Reduction in Central Asia."

123. Choffnes, "Germs on the Loose," 60.

124. Kenneth L. Calder, "Preliminary Discussion of Methods for Calculating Munition Expenditures, with Special Reference to the St. Jo Program," Camp Detrick memo, Aug. 11, 1954, pp. 74, 76; Leonard A. Cole, *Clouds of Secrecy: the Army's Germ Warfare Tests over Populated Areas* (New York: Rowman and Littlefield, 1988); Miller et al., *Germs,* 42–43.

125. Cole, *Clouds of Secrecy,* 64.

126. Department of Defense, "Biological Defense Program," May 1986, excerpted in Cole, *Clouds of Secrecy,* appendix 4, 176.

127. Martin Hugh-Jones, "Wickham Steed and German Biological Warfare Research," *Intelligence and National Security* 7 (1992): 379–402.

128. Department of Defense, "Biological Defense Program," 179.

129. Balmer, "The UK Biological Weapons Program," 77–78, Alibek, *Biohazard,* 234.

Chapter 6. Detection and Verification

Epigraph: John Ezzell quoted in Marilyn Thompson, *The Killer Strain: Anthrax and a Government Exposed* (New York: Harper-Collins, 2003), 120.

1. Thompson, *The Killer Strain.*

2. Ibid., 120.

3. For our purposes, *premeditation* includes (1) having the time and capacity to create the agent and plan an attack, and (2) having the intent to frighten, debilitate, or kill victims with the biological weapon. When victims are other than the intended targets, Anglo-American criminal law still ascribes culpability through the doctrine of transferred intent. See American Law Institute, *Model Penal Code and Commentaries,* sec. 2.02(6) (Philadelphia: ALI, 1980). A vast legal literature on premeditation, especially in cases of murder, exists; a good starting point is Matthew A. Pauley, "Murder by Premeditation," *American Criminal Law Review* 36 (March 1999): 145–70.

4. Bioterrorism is differentiated from biocriminality by the intent of the perpetrator(s). The motive for bioterrorism is political, ideological, or religious; biocriminality is often perpetrated by individuals for purposes of revenge or financial gain. See Mark Wheelis, "Bioterrorism," *Pestilence, Pandemics, and Plagues,* ed. Joseph P. Byrne (Westport, CT: Greenwood Press, 2008), 53–56.

5. The first reports of the outbreak appeared in the news media: *Posev* (Frankfurt, Germany) 1, 7 (1980): 4; Bernard Gwertzman, *New York Times,* March 19, 1980, 1. The contaminated-meat explanation was published in articles in Soviet veterinary and medical journals. See I. S. Bezdenezhnykh and V. N. Nikiforov, *Zhurnal Mikrobiologii, Epidemiologii, i Immunobiologii,* no. 5 (1980): 1–11; *Veterinariya,* no. 10 (1980): 3; *Chelovek i Zakon* 117, 9 (1980): 70.

6. Elisa D. Harris, "Sverdlovsk and Yellow Rain: Two Cases of Soviet Noncompliance?" *International Security* 11 (Spring 1987): 41–95, see 45.

7. N. Zhenova, *Literatumaya Gazeta* (Moscow), Aug. 22, 1990, 12; ibid., Oct. 2, 1991, 6; S. Parfenov, *Rodina* (Moscow), May 1990, 21; V. Chelikov, *Komsomolskaya Pravda* (Moscow), Nov. 20, 1991, 4.

8. See for example R. Jeffrey Smith, "Soviet Anthrax Explanation Debunked," *Science* 209 (July 18, 1980): 375.

9. D. Muratov, Yuri Sorokin, V. Fronin, *Komsomolskaya Pravda* (Moscow), May 27, 1992, 2, described in Matthew Meselson, Jeanne Guillemin, Martin Hugh-Jones, Alexander Langmuir, Ilona Popova, Alexis Shelokov, and Olga Yampolskaya, "The Sverdlovsk Anthrax Outbreak of 1979," *Science* 266 (Nov. 18, 1994): 1202–1208, see 1203.

10. Ken Alibek, with Stephen Handelman, *Biohazard: The Chilling True Story of the Largest Covert Biological Weapons Program in the World—Told from Inside by the Man who Ran It* (New York: Dell Publishing, 1999), 84, 99.

11. Fred Guterl and Eve Conant, "In the Germ Labs: The Former Soviet Union Had Huge Stocks of Biological Agents. Assessing the Real Risk," *Newsweek* 139, 8 (Feb. 25, 2002): 26–28.

12. Meselson et al., "The Sverdlovsk Anthrax Outbreak," 1203.

13. Jeanne Guillemin, *Anthrax: The Investigation of a Deadly Outbreak* (Berkeley: University of California Press, 1999).

14. Ibid., 86–91.

15. Meselson et al., "The Sverdlovsk Anthrax Outbreak," 1206.

16. Guillemin, *Anthrax,* 190.

17. Jeanne Guillemin, "The 1979 Anthrax Epidemic in the USSR: Applied Science and Political Controversy," *Proceedings of the American Philosophical Society* 146, 1 (March 2002): 18–36, map on 24; Meselson quoted in Bob Coen and Eric Nadler, *Dead Silence: Fear and Terror on the Anthrax Trail* (Berkeley, CA: Counterpoint, Transformer Films, 2009), 114.

18. Coen and Nadler, *Dead Silence,* 110.

19. D. W. Henderson, S. Peacock and F. C. Belton, "Observations on the Prophylaxis of Experimental Pulmonary Anthrax in the Monkey," *Journal of Hygiene* 54 (1956): 28–36; H. N. Glassman, "Discussion," *Bacteriologic Review* 30 (1966): 657–59; Guillemin, "The 1979 Anthrax Epidemic," 27.

20. Philip S. Brachman, Stanley A. Plotkin, Forrest H. Bumford, and Mary M. Atchison, "An Epidemic of Inhalation Anthrax: The First in the Twentieth Century," *American Journal of Hygiene* 72 (July 1960): 6–23; Charles M. Dahlgren, Lee M. Buchanan, Herbert M. Decker, Samuel W. Freed, Charles R. Phillips, and Philip S. Brachman, "*Bacillus anthracis* aerosols in Goat Hair Processing Mills," *American Journal of Hygiene* 72 (July 1960): 24–31.

21. Guillemin, *Anthrax,* 161.

22. Ibid., 220–26, quotation on 225.

23. P. J. Jackson, M. E. Hugh-Jones, D. M. Adair, G. Green, K. K. Hill, C. R. Kuske, L. M. Grinberg, F. A. Abramova, and P. Keim, "PCR Analysis of Tissue Samples from the 1979 Sverdlovsk Anthrax Victims: The Presence of Multiple *Bacillus anthracis* Strains in Different Victims," *Proceedings of the National Academy of Sciences USA* 95 (1998): 1224–29.

24. The verification of natural versus artificial outbreaks was a central problem in biological weapons control. See Stephen Meyer, "Verification and Risk in Arms Control," *International Security* 8 (Spring 1984): 115–16. As Michael D. Gordin has written, the need for verification "relied heavily on an oversimplified view of science that assumed that a single interpretation of evidence was always possible." See Gordin, "The Anthrax Solution: The Sverdlovsk Incident and the Resolution of a Biological Weapons Controversy," *Journal of History of Biology* 30 (1997): 441–80, see 472 and 477.

25. Gordin, "The Anthrax Solution," 480.

26. Guillemin, *Anthrax*, 262.

27. Ibid., 238 (Guillemin quotation) and 161 (Meselson quotation).

28. Martin Hugh-Jones, "Distinguishing Between Natural and Unnatural Outbreaks of Animal Diseases," *Revue Scientifique et Technique (Office International des Epizooties)* 25, 1 (2006): 173–86, quotation on 173.

29. Guillemin, *Anthrax*, 243.

30. P. N. Burgasov, B. L Cherkasskiy, L. M. Marchuk, and Y. F. Shcherbak, *Anthrax [Sibirskaya Yazva]* (Moscow: Editions Medicine, 1970; trans. Foreign Broadcast Information Service, Washington, D.C., 1981), quotation on 3–4 (also cited in Guillemin, *Anthrax,* 33); statistics in Guillemin, *Anthrax,* 33–34.

31. Leonard A. Cole, *The Anthrax Letters: A Medical Detective Story* (Washington, D.C.: Joseph Henry Press, 1993), 48. See also Thompson, *The Killer Strain.*

32. An excellent synopsis of events early in the investigation can be found in Center for Counterproliferation Research, "Anthrax in America: A Chronology and Analysis of the Fall 2001 Attacks" (Working paper, National Defense University, Washington, D.C., 2002).

33. Roger Breeze, Bruce Budowle, and Steven E. Schutzer, *Microbial Forensics* (New York: Academic Press, 2005).

34. Randall S. Murch, "History, Strategy and Future of Microbial Forensics," lecture presented at Microbial Forensics: The Nexus between Law Enforcement and Public Health and Science, AAS-NAS Workshop, Jan. 24, 2008.

35. Guillemin, *Anthrax,* 161, 166.

36. J. A. Jerrigan, D. S. Stephens, D. A. Ashford, C. Omenaca, M. S. Topiel, M. Galbraith, M. Tapper, T. L. Fisk, S. Zaki, T. Popovic, R. F. Meyer, C. P. Quinn, S. A. Harper, S. K. Fridkin, J. J. Sejvar, C. W. Shepard, M. McConnell, J. Guarnar, W. J. Shieh, J. M. Malecki, J. L. Gerberding, J. M. Hughes, B. A. Perkins, and Anthrax Bioterrorism Investigation Team, "Bioterrorism-Related Inhalational Anthrax: The First Ten Cases Reported in the United States," *Emerging Infectious Dis-*

eases 7, 6 (Nov.–Dec. 2001): 933–44; H. Clifford Lane and Anthony S. Fauci, "Bioterrorism on the Home Front: A New Challenge for American Medicine," *Journal of the American Medical Association* 286, 20 (Nov. 28, 2001): 2595–97.

37. "Anthrax Scares Hit Postal Centers in New Zealand and Australia," *Independent*, Oct. 17, 2001, A1.

38. "Anthrax Scares Mount Across UK," *Guardian*, Oct. 16, 2001, 1.

39. "Postman Jailed for Racist Hate Mail Campaign," *Guardian*, June 13, 2008, www.guardian.co.uk/uk/2008/jun/13/ukcrime4 (accessed July 5, 2008).

40. The exact nature of this process and of the additive to the spores remains controversial, with a confusing array of contradictory reports published in various literatures. See U.S. House of Representatives, Committee on Homeland Security, "Engineering Bio-terror Agents: Lessons from the Offensive U.S. and Russian Biological Weapons Programs," *Serial No. 109-30* (Washington, D.C.: U.S. Government Printing Office, July 13, 2005); Dany Shoham and Stuart M. Jacobsen, "Technical Intelligence in Retrospect: The 2001 Anthrax Letters Powder," *International Journal of Intelligence and CounterIntelligence* 20 (March 2007): 79–105; Kay A. Mereish, "Unsupported Conclusions on the *Bacillus anthracis* Spores," *Applied and Environmental Microbiology* 73, 15 (2007): 5074.

41. In late 1990, during the presidential administration of George H. W. Bush, the Iraqi army had invaded neighboring Kuwait. A coalition of thirty-two nations (including the United States, Britain, Egypt, France, and Saudi Arabia) then responded in kind in January 1991, pushed the Iraqis out of Kuwait, and invaded Iraq. Rumors circulated that the Iraqis had biological and chemical weapons programs, so 150,000 American military personnel were given immunizations against anthrax. In 2001, during the George W. Bush administration, this history influenced the initial conclusion of American authorities.

42. Thompson, *The Killer Strain*, 49.

43. "Ames" is a misnomer; the bacilli actually came from a sick cow in Texas in 1981. This fact was uncovered during the Amerithrax investigation.

44. Peter J. Boyer, "The Ames Strain," New Yorker, Nov. 12, 2001. The destruction of the spores, although urgent at the time, meant that a great deal of the genetic historical record of *B. anthracis* in the United States was lost.

45. Thompson, *The Killer Strain*, 199.

46. Cole, *The Anthrax Letters*, 198.

47. Lois R. Ember, "Anthrax Sleuthing: Science Aids a Nettlesome FBI Criminal Probe," *Chemical and Engineering News* 84, 49 (Dec. 2006): 47–54, see 51–52.

48. T. D. Read, S. L. Salzberg, M. Pop, M. Shumway, L. Umayam, L. Jiang, E. Holtzapple, J. Busch, K. Smith, J. Schupp, D. Solomon, P. Keim, C. Fraser, "Comparative Genome Sequencing for Discovery of Novel Polymorphisms in *Bacillus anthracis*," *Science* 296, 5575 (May 2002): 2028–33.

49. Center for Counterproliferation Research, "Anthrax in America," 47.

50. W. J. Broad and J. Miller, "A Nation Challenged: The Investigation; U.S.

Recently Produced Anthrax in a Highly Lethal Powder Form," *New York Times,* Dec. 13, 2001, A1.

51. See for example Gary Matsumoto, *Vaccine A: The Covert Government Experiment That's Killing Our Soldiers and Why GI's Are Only the First Victims* (New York: Basic Books, 2004); Matsumoto, "The Pentagon's Toxic Secret," *Vanity Fair* (May 1999): 82–98.

52. Leonard A. Cole, "Anthrax Hoaxes: Hot New Hobby?" *Bulletin of the Atomic Scientists* 55, no. 4 (July 1, 1999): 7–11, see 9.

53. Michael R. Gordon, "U.S. Says It Found Qaeda Lab Being Built to Produce Anthrax," *New York Times,* March 23, 2002, A1.

54. Sheryl Gay Stolberg and David E. Rosenbaum, "U.S. Will Offer Anthrax Shots for Thousands," *New York Times,* Dec. 19, 2001, A1.

55. Wendy Orent, "Cattle Call," *New Republic* 225 (Nov. 12, 2001): 16–18, quotation on 18.

56. The United States Department of Justice officially closed the Amerithrax investigation on February 19, 2010.

57. Thompson, *The Killer Strain,* 34.

58. Jeanne Guillemin, *Biological Weapons: From the Invention of State-Sponsored Programs to Contemporary Bioterrorism* (New York: Columbia University Press, 2005), 167.

59. Cole, *The Anthrax Letters,* 141.

60. John Ezzell, quoted in Thompson, *The Killer Strain,* 209.

61. This technique is *polymerase chain reaction,* or PCR.

62. Thompson, *The Killer Strain,* 38–39.

63. Rosie Mestel, "Anthrax's Dogged Detective," *Los Angeles Times,* Dec. 17, 2001.

64. Matthew N. Van Ert, W. Ryan Easterday, Lynn Y. Huynh, Richard T. Okinaka, Martin E. Hugh-Jones, Jacques Ravel, Shaylan R. Zanecki, Talima Pearson, Tatum S. Simonson, Jana M. U'Ren, Sergey M. Kachur, Rebecca R. Leadem-Dougherty, Shane D. Rhoton, Guenevier Zinser, Jason Farlow, Pamala R. Coker, Kimothy L. Smith, Bingxiang Wang, Leo J. Kenefic, Claire M. Fraser-Liggett, David M. Wagner, and Paul Keim, "Global Genetic Population Structure of *Bacillus anthracis,*" *PLoS One* 2, 5 (May 2007): e461, 1–10, see 7; A. Fasanella, M. E. Van Ert, S. A. Altamura, G. Garofolo, C. Buonavoglia, P. Keim, M. C. Chu, and K. L. Gage, "Molecular Diversity of *Bacillus anthracis* in Italy," *Journal of Clinical Microbiology* 43 (2005): 3398–3401; J. L. Lowell, D. M. Wagner, B. Atshabar, M. F. Antolin, A. J. Vogler, et al., "Identifying Sources of Human Exposure to Plague," *Journal of Clinical Microbiology* 43 (2005): 650–56; D. T. Cheung, K. M. Kam, K. L. Hau, T. K. Au, C. K. Marston, "Characterization of a *Bacillus anthracis* Isolate Causing a Rare Case of Fatal Anthrax in a 2-year-old Boy from Hong Kong," *Journal of Clinical Microbiology* 43 (2005): 1992–94.

65. Abigail Salyers quoted in Thompson, *The Killer Strain,* 201.

66. Susan Candiotti, "Anthrax Terror Remains a Mystery," *CNN,* March 27, 2002, http://archives.cnn.com/2002/US/03/26/anthrax.investigation (accessed Aug. 7, 2009).

67. John Ashcroft, on *The Early Show,* CBS, Aug. 6, 2002.

68. Thompson, *The Killer Strain,* 189–97, quotation on 189; Cole, *The Anthrax Letters,* 192–95.

69. Hatfill won a $4.6 million judgment in the summer of 2008. See Scott Shane and Eric Lichtblau, "Scientist is Paid Millions by U.S. in Anthrax Suit," *New York Times,* June 28, 2008, www.nytimes.com/2008/06/28/washington/28hatfill.html?_r=1&fta=y&oref=slogin (accessed July 1, 2008).

70. Scott Shane, " 'Amerithrax' Easy to Make," *San Francisco Chronicle,* April 11, 2003, A-4.

71. Along with the mistaken accusation of Hatfill, the investigation has been criticized for inefficiency and inaccuracies (see Thompson, *The Killer Strain,* for example).

72. Paul Keim, *Microbial Forensics: A Scientific Assessment* (Washington, D.C.: American Academy of Microbiology, 2003), 5.

73. Nicholas Wade, "A Trained Eye Finally Solved Anthrax Puzzle Through 'Morphing' Samples," *New York Times* Aug. 21, 2008, A1 (national edition).

74. Matthew N. Van Ert, W. Ryan Easterday, Tatum S. Simonson, Jana M. U'Ren, Talima Pearson, Leo J. Kenefic, Joseph D. Busch, Lynn Y. Huynh, Megan Dukerich, Carla B. Trim, Jodi Beaudry, Amy Welty-Bernard, Timothy Read, Claire M. Fraser, Jacques Ravel, Paul Keim, "Strain-Specific Single-Nucleotide Polymorphism Assays for the *Bacillus anthracis* Ames Strain," *Journal of Clinical Microbiology* 45 (January 2007): 47–53, see 52.

75. Wade, "A Trained Eye," A14.

76. Thomas F. Dellaferra, Department of Justice, search warrant affidavit 07-529-M01, Oct. 31, 2007; Scott Shane and Eric Lichtblau, "Scientist's Suicide is Linked to Anthrax Inquiry," *New York Times,* Aug. 2, 2008, A1.

77. See for example Department of Justice, search warrant returns #08-430, 07-524-M-01, and 07-529-M-01, Federal Bureau of Investigation, www.usdoj.gov/amerithrax (accessed Aug. 7, 2008); William J. Broad and Scott Shane, "For Suspects, Anthrax Case Had Big Costs," *New York Times,* Aug. 10, 2008, A1. Bruce Ivins' co-author of the 1983 landmark paper about *B. anthracis*' plasmids, Perry Mikesell, also came under suspicion and surveillance. Mikesell began drinking heavily and died in October 2002.

78. Shane and Lichtblau, "Scientist's Suicide"; Scott Shane and Eric Lichtblau, "FBI Presents Anthrax Case, Saying Scientist Acted Alone," *New York Times,* Aug. 7, 2008, A1.

79. Shane and Lichtblau, "Scientist's Suicide"; Shane and Lichtblau, "FBI Presents Anthrax Case."

80. Dellaferra, search warrant affidavit 07-529-M01; Michael Stebbins, "Bush

Administration Cancels Anthrax Vaccine Contract," Dec. 20, 2006, Federation of American Scientists Strategic Security Blog, www.fas.org/blog/ssp/2006/12/bush_administration_cancels_an.php (accessed Aug 5, 2008).

81. Michael Isikoff, "Justice Said the Evidence against the Late Ivins Is Strong Enough to 'Prove His Guilt,'" *Newsweek*, Aug. 9, 2008, www.newsweek.com/id/151784 (accessed Aug. 10, 2008).

82. Quote in David Willman, "Anthrax Scientist Bruce Ivins Stood to Benefit from a Panic," *Los Angeles Times*, Aug. 2, 2008.

83. Eric Lichtblau and Scott Shane, "Anthrax-Case Affidavits Add to Bizarre Portrait of Suspect," *New York Times*, Sept. 25, 2008, A22 (East Coast late edition).

84. Scott Shane, "Critics of Anthrax Inquiry Seek an Independent Review," *New York Times*, Sept. 24, 2008, A17. For more on congressional skepticism, see the transcript of attorney and blogger Glenn Greenwald's radio interview with New Jersey representative Rush Holt, chairman of the House's Select Intelligence Oversight Panel, Aug. 5, 2008, www.salon.com/opinion/greenwald/radio/2008/08/05/holt (accessed Aug. 7, 2008). See also Amy Goodman's interviews of Greenwald and Meryl Nass, "Anthrax Mystery: Questions Raised over Whether Government Is Framing Dead Army Scientist for 2001 Attacks," *Democracy Now* radio/television program, Aug. 4, 2008, transcript at: www.democracynow.org/2008/8/4/anthrax (accessed on Aug. 5, 2008). Physician and anthrax expert Meryl Nass's opinions were published on the Internet at anthraxvaccine.blogspot.com.

85. Eric Lichtblau, "Independent Review Set On F.B.I. Anthrax Inquiry," *New York Times*, Sept. 17, 2008, A19; "FBI Call on NAS to Study Anthrax Case," *Physics Today*, News Picks, Oct. 6, 2008, http://blogs.physicstoday.org/newspicks/2008/10/fbi_call_on_nas_to_study_anthr.html (accessed Nov. 1, 2008).

86. Eric Lipton and Scott Shane, "Anthrax Case Renews Questions on Bioterror Effort and Safety," *New York Times*, Aug. 3, 2008, A1.

87. Guillemin, *Anthrax*, 246.

Epilogue

Epigraph: Theodor Rosebury, *Peace or Pestilence: Biological Warfare and How to Avoid It* (New York: McGraw-Hill, 1949), 197.

1. Rosebury, *Peace or Pestilence*, 35, 62, 63.

2. For recent treatments of the concepts of first and second nature, see William Cronon, *Nature's Metropolis: Chicago and the Great West* (New York: W.W. Norton, 1991), xix; Reiner Grundmann, "The Ecological Challenge to Marxism," *New Left Review* 187 (1991): 103–20.

3. Macfarlane Burnet and David O. White, *Natural History of Infectious Disease*, 4th ed. (Cambridge: Cambridge University Press, 1974), 267.

4. Ibid., x.

5. Robert Wright, "Be Very Afraid: Nukes, Nerve Gas and Anthrax Spores," *New Republic* 212 (May 1, 1995): 19–27, quotation on 26.

6. Keim quoted in Rosie Mestel, "Anthrax's Dogged Detective," *Los Angeles Times,* Dec. 17, 2001.

7. Rosebury, *Peace or Pestilence,* 174.

8. Guillemin, *Anthrax,* 228.

9. Rosebury, *Peace or Pestilence,* 182.

10. Ann Finkbeiner, *The Jasons: The Secret History of Science's Postwar Elite* (New York: Viking, 2006), 218. I thank Jacqueline Wehmueller for recommending this study.

11. Ibid., 227.

12. Ibid., 222.

13. See for example David Willman, "Anthrax Scientist Bruce Ivins Stood to Benefit from a Panic," *Los Angeles Times,* Aug. 2, 2008.

14. This line appeared in at least two places. See R. W. Apple, Jr., "City of Power, City of Fear," *New York Times,* Oct. 18, 2001, A1; Marianne Szegedy-Maszak, "Infected Letters Spread a Contagion of Fear that Goes Far Beyond Germs," *U.S. News and World Report* 131, 19 (Nov. 5, 2001): 47.

15. Apple, "City of Power, City of Fear."

16. Editorial, "How Soccer Moms Became Security Moms," *Time Magazine* 161, 7 (Feb. 17, 2001): 23.

17. Michael Kinsley, "Be a Patriot. Don't Hoard Cipro!" *Time Magazine* 158, 19 (Oct. 29, 2001): 73.

18. Michael Lemonick, "How to Deal with Anxiety," *Time Magazine* 158, 19 (Oct. 29, 2001): 91; Bruce Shapiro, "Anthrax Anxiety," *Nation* 273, 15 (Nov. 5, 2001): 4–5.

19. For the record, *B. anthracis* spores do not have a distinctive smell.

20. René Dubos, *The Mirage of Health: Utopias, Progress, and Biological Change* (New York: Harper and Brothers, 1959), 1.

INDEX